ARCHETYPES AND ANCESTORS

Palaeontology in Victorian London
1850–1875

Richard Owen in 1872. (From *The Illustrated London News*, 3 February 1872, 117.)

ARCHETYPES AND ANCESTORS

Palaeontology in
Victorian London
1850–1875

ADRIAN DESMOND

The University of Chicago Press

The University of Chicago Press, Chicago 60637
Frederick Muller Limited, Dataday House, London SW19 7JZ
© 1982 by Adrian Desmond
All rights reserved. Published 1982
University of Chicago Press edition 1984
Printed in the United States of America

84 85 86 87 88 89 90 91 92 93   5 4 3 2 1

Library of Congress Cataloging in Publication Data

Desmond, Adrian J., 1947–
 Archetypes and ancestors.

 Bibliography: p.
 Includes index.
 1. Paleontology—England—London—History.   I. Title.
QE705.G7D47   1984   560'.9421   83-18104
ISBN 0-226-14343-0
ISBN 0-226-14344-9 (pbk.)

Contents

Acknowledgments

In the six years spent researching and writing *Archetypes and Ancestors*, I have naturally chalked up debts of gratitude to a great many individuals and institutions. First and foremost, I would like to thank Nellie Flexner, J. A. Cowie, and Barbara Desmond for checking an endless succession of typescripts, and for doing more towards improving my grammar than merely curing my occasional attacks of Victorian punctuation. Bill Bynum, Mike Bartholomew, Jim Secord, and the late Dov Ospovat cast a historical eye over the chapters; while Chris Duffin, Kevin Padian, Mike Benton, and Pamela Gill looked over the palaeontology. Many were good enough to send me unpublished or in-press material; in this respect, I must also thank Steven Shapin for letting me preview his forthcoming paper, and Jon Hodge for kindly forwarding a transcription of an unpublished Richard Owen letter. At University College London, Richard Freeman talked over the suspected homosexuality of the first professor of zoology, Robert Grant; and Pamela Robinson discussed Grant's successor, Ray Lankester.

Jeanne Pingree at the Imperial College Archives and M. J. Rowlands of the British Museum (Natural History) made it possible for me to study and transcribe the correspondence of T. H. Huxley and Richard Owen respectively. In addition, the following bodies generously granted permission to quote from manuscript material in their care: the President and Council of the Royal College of Surgeons of England, British Library, University College London Library, the Syndics of Cambridge University Library (with especial thanks to P. J. Gautrey), and the Alexander Turnbull Library,

Wellington, New Zealand. The Royal Institution, Wellcome Institute for the History of Medicine, Edinburgh University Library, Zoological Society of London, and Royal Society allowed me to read autograph letters, journals, and referees' reports. Patricia Methven, Archivist at King's College, London, supplied information on Martin Duncan; N. H. Robinson sent me details of Royal Society referees' reports; and Elizabeth Allen, Curator of the Hunterian Museum, made available Owen's interleaved catalogues and attempted to trace his casts of the fossil amphibian *Archegosaurus*.

My research was primarily carried out in the D. M. S. Watson Natural Sciences Library at University College London, among the splendid palaeontological reprint and book collections, and in the Medical Sciences Library. As always it is a pleasure to acknowledge the help of librarians Joan Nash and Susan Gove, who succeeded in meeting my constant demands for obscure material. The book was conceived and begun in 1975, while I was on leave from Harvard, and this brings me to a far more general debt – to Everett Mendelsohn, for introducing me to sociological techniques in the history of science, and to Stephen Jay Gould, for his encouragement in those distant days, and for sharing his expertise on evolutionary matters, past and present.

ARCHETYPES
AND
ANCESTORS

Introduction

Mid-Victorian London was the Western world's most populous city, with almost 2.7 million people at the time of the 1851 census, and 3.9 million twenty years later. The metropolis was already suffering from traffic jams and 'rush hours' by the sixties, and pea-soupers worse than anyone could remember – "hideous, vicious, cruel, and above all overwhelming," exclaimed Henry James on a visit in 1868, "a dreadful, delightful city."[1] But by the end of our period it *was* recognisably London. The creation of a Metropolitan Board of Works in 1855 effectively unified local government. The Board completed the great sewerage systems, built the Embankment from Chelsea to Blackfriars, freed many toll bridges over the Thames, cut new roads and established the fire brigade. In his classic *Portrait of an Age*, G. M. Young rhapsodised over this conversion of "the vast and shapeless city which Dickens knew – fog-bound and fever-haunted, brooding over its dark, mysterious river – into the imperial capital, of Whitehall, the Thames Embankment and South Kensington".[2] The frenetic Western metropolis had become the nerve centre of an industrial empire and, judging by the reaction of foreigners, the wonder of its age.

Historical works are sometimes hard-put to justify narrowing their focus, and rightly so if this means losing the larger picture. My problem is magnified, since I intend to concentrate not only on three decades in a bustling century, but on a single city at its centre – London. Couple this with the fact that I am primarily interested in the development of a science, palaeontology, and one might be excused a certain trepidation. Actually the outcome is not so esoteric after all. To the obvious question, Why London and why those dates, 1850–1875? I would answer that this narrowing is crucial if we want to relate the major social and cultural changes to the 'deep structure' of our science.

Victoria's reign was, of course, immensely long. It opens "with bishops in cauliflower wigs and the great ones of the world driving in coaches with footmen behind, it ends with expensive people driving in motor cars and a leader of the House of Commons who rode a bicycle".[3] And behind the revolution in transport lay a more profound social transformation, with the rise of suffrage, democracy, the middle classes and organised labour, all of which affected in an indirect but no less real way the passage from a Georgian Paleyite world deliberating on the power, wisdom, and goodness of God, to one with a professionally-organised bourgeoisie probing the genetic basis of life. Those who single out what Kitson Clark calls the "High Noon of Victorianism", generally do so, in the first instance, on economic grounds. Wedged between the hungry forties and socialist eighties, this period of calm has been likened to a sunny afternoon sandwiched between a stormy morning and menacing night (a metaphor whose worth really depends upon one's political bent). By and large, it was a time of peace and prosperity for most sections of the community, even, according to Kitson Clark, of "secure opulence". The passing of the Chartist storms in 1848, with their attendant hopes and fears, paved the way for the eminently respectable age of Palmerston, Disraeli, and Gladstone. Workmen at last hoped to better their lot within the system, and even though the interlude lacked any major reforms until 1867 because of effete government, they were not disappointed (although the poorest of them would hardly have noticed their good fortune). The boom years lasted two decades, a generation of the Pax Britannica and sterling imperialism, the end roughly signalled by the Franco-Prussian War and the Depression of 1873, when British manufacturers finally began to face stiff competition from Western Europe and the United States. Needless to say, such a synopsis does an injustice to the fine texture of "High Noon" history, smoothing the wrinkles and side-stepping the minor crises, crushing and compacting events into a few lines. Nonetheless, as a first approximation the image was true enough for W. L. Burn to title his study of the mid-Victorian period *The Age of Equipoise*.

We begin shortly before the opening of that national showpiece, the Great Exhibition of 1851. This symbolised the culmination of the romantic age as much as the ascendancy of the utilitarian, commercial middle-classes.[4] So it is no coincidence that the transcendental anatomy of the great Richard Owen (himself the designer of dinosaurs for the new Exhibition grounds at Sydenham in 1854) was ceremoniously jettisoned in the fifties by the 'young guard', Herbert Spencer, T. H. Huxley, and their circle, whose 'proletarian' science prepared the ground for Darwin's *Origin of Species* (1859). Owen's Platonic and other-worldly Archetype, the ideal invocation of vertebrate life, was ignominiously traded for a common-or-garden ancestor. And the tenacious middle-class popularisers made sure it was seen to be so by the masses. Huxley and the physicist John Tyndall, with their hardnosed 'mercantile' and secular approach, deliberately wooed the wage-earners, the sort who flocked in their millions to gawp at the antediluvian monsters at Crystal Palace. The naturalism the young guard espoused "was the ideology of rapidly ascending new professional groups in an industrial society", say sociologists Barry Barnes and Steven Shapin – a means of legitimising a new secular scientific authority. And in this they were sustained by an increasingly-powerful bourgeoisie "whose commercial and industrial activities set them in opposition to the landed interest and its traditional minions."[5] These immense changes in the social foundation of science were reflected in the quick capitulation of key posts within universities and learned societies to the new man. Their working assumption was nature's uniformity and they argued persuasively for unbroken physical causation, and like a number of their Christian colleagues supported Darwin and the higher biblical criticism of *Essays and Reviews*.

New institutions bore witness to the social upheaval, foremost being the Museum of Practical Geology – according to Bernard Becker's *Scientific London* "the best example in this country of what a museum ought to be."[6] The building was rushed to completion in time for the opening of the Great Exhibition in Hyde Park. It stretched from Jermyn Street to Piccadilly, and was proudly proclaimed as the first in Britain designed specifically as a scientific institution. Attached to it were the research laboratories and lecture halls of the Government School of Mines. Pre-eminently utilitarian, the new School – with its "factory discipline mentality", as Roy Porter recently put it – was designed not only to train students and technologists, but to provide popular lectures to working men "interested in the progress of science and its application to manufactures",[7] as well as analyse minerals for moderate fees and construct geological maps. Here Huxley was to teach, as were a number of others among the young guard, including the botanist J. D. Hooker and Tyndall for a time – although notoriously Owen's attempt in 1857 to deliver a course on fossil vertebrates at the school caused such a furore that Huxley formally severed all relations with him (Chapter 1).

By our period's end, the "late afternoon" (the recurrent metaphor is this time G. M. Young's), most younger London scientists called themselves "evolutionists". The extent to which the word was becoming fashionable was brought home to the incorrigible Owen, when he gave a lecture on crocodile modification to the Geological Society in 1878 which avoided it.[8] On the whole, though, palaeontology in the Depression was losing its "polemical element" – and with it popular appeal, as Becker discovered when he dropped in on one of Martin Duncan's afternoon talks at the Royal Institution in 1874. Attendance was poor, and he concluded that without the "controversial spice", the kind that Owen dished up at the time of the *Origin* and *Essays and Reviews*, the subject forfeited its "whilom attractiveness". Then again, it is just as likely that the emergence of career-orientated geologists like Duncan, a coral expert from King's, and the narrow specialisation which was coming to characterise the science, left amateurs and dabblers feeling helpless and unwelcome, and drove popular audiences in search of newer pursuits.

Whether the period saw an increase in the number of fossils shipped to London I do not know; certainly crates seemed to be arriving in Bloomsbury in ever increasing numbers from the colonies (particularly South Africa, New Zealand, and Australia), and my impression is that the "Coal age" vertebrates deposited in Jermyn Street increased considerably, as one might have expected given Britain's predominant mining interests. When Darwin joyously exclaimed to Wallace in 1876 "What progress Palaeontology has made during the last twenty years!"[9] he mostly had in mind new evidence for geographical distribution and probably the use of key fossils to patch up the geological record. But when the visiting Philadelphian Edward Drinker Cope two years later called London a palaeontological Babylon he was referring to the sheer quantity of fossils. The national collections evidently surpassed those in Germany early on, for in 1860 the Swiss histologist Albert von Kölliker warned Huxley of an impending visit "as I am very needful of information, with regard to fossils, and as I can get this nowhere better than in London."[10]

It might be guessed that I have no great love for merely cataloguing collections. Nor am I easy about the sort of inductivist approach which sees changing theories of life depend necessarily on new fossil finds, or which views historical progression solely through the increasing knowledge of fossils themselves. The advantage of Martin Rudwick's *Meaning of Fossils* was that it taught us to place contemporary interpretations within their philosophical context, and when we likewise bring social concerns to the fore the way is open for a broader analysis of Victorian trends that avoids the Whiggishness of older accounts.[11] Rudwick's compass was wide, from the Renaissance to Victoria's reign; my scope is a mere thirty years or so, which will allow me to delve more deeply into specific controversies and their foundations. The practical upshot of this approach is – to take an example – that I have not fêted *Archaeopteryx*, nor accorded it the decisive role so often encountered in partisan histories (too many tracts for their times have plundered history to support a modern evolutionary view). The situation in Darwin's day was inestimably more complex, and I shall argue that the importance of *Archaeopteryx* lay in its use to rival social factions with conflicting strategies: in short, it was seen through different eyes and satisfactorily accommodated in a number of divergent ways. Not only could the interpretation of individual fossils vary with covert social factors, but in some cases even the restored fossil animals (such as Owen's quadrupedal dinosaurs at Crystal Palace and Huxley's bird-like rivals) were built to 'social' specifications, i.e. they were designed for *use*, to support contrasting scientific claims. They betrayed their builders' intent and so provide us with a perfect case study in cultural impact. Whether or not I am right in detail, this surely is a more equitable way of tackling Victorian palaeontology – not as following a 'true' or 'erroneous' path, or failing miserably to comprehend the nature of long-extinct life because the twentieth century has arrived at different conclusions (such unbridled Whiggism is anyway unfashionable in science history[12]), but as demonstrating that contemporary science was in accord with the cultural norms of the age.

For this reason, and despite the obvious dangers, I have found it constructive to focus on the shift in power from the romantic Owen to up-and-coming "professionalisers" like Huxley and his protégé W. H. Flower, who began their careers in the fifties. Despite the comparative affluence of the period, there was little money and less prestige in 'pure' research at this time. Until the shift to collectivism in the seventies, which manifested throughout society, from political to scientific organisation, there was no centrally-funded body scientific as such (unless it be the Royal Society or British Association). With individualism the dominant creed (and vigorously supported by Spencer), scientists found Treasury funding notoriously difficult to obtain. Flower once complained that he "could make more thousands as a surgeon than he ever would hundreds as a man of science."[13] The "professionalisers", in tune with the times, set out to change all this, campaigning for Government support and State funding. They were almost all Londoners by birth or choice; something certainly true of that "open conspiracy", the x Club (a group of Huxley's intimates, devoted "to science, pure and free, untrammelled by religious dogma",[14] who dined together once a month at St George's Hotel in Albemarle Street). Huxley was born above a butcher's shop in Ealing, Hooker was Assistant and afterwards Director at Kew Gardens, the Orangeman Tyndall had stepped into Faraday's shoes at the Royal Institution. Spencer was an individualist in more senses than one, reflected in the fact that he worked for a while on *The Economist*, but he did share the group's naturalistic and secular leanings. It was difficult to devote oneself full-time to research in the provinces, at least without private means, even supposing museum materials to exist. One might, at most, hope to live off writing or a profitable practice and take up fossils as a hobby (although the toll this double life took on Gideon Mantell's health and marriage should have been a caution to all). Even so, being out of personal touch, provincial scientists might have found professional recognition much harder to come by. Take the case of the young Unitarian physiologist W. B. Carpenter. He felt hopelessly isolated in Bristol. "Living as I do," he lamented in a letter to Owen in 1842, "so completely out of the way of knowing what is being done in Science, except through the ordinary Journals, I am always uncertain if I am really working to any advantage".[15] As a result Carpenter uprooted his family and made the uncertain journey to London, where he became eminently successful, befriending Huxley, teaching at University College, and in 1856 being elected Registrar of the University.

Carpenter's support for Huxley in the fifties and sixties brings me to my last point. Although *x* Club agitation was primarily for secular and naturalistic ends, it enjoyed the backing of many young Christian biologists in the city. Carpenter for one, the Methodist anatomist and proverbial boffin W. K. Parker for another, and of course the Broad Churchman Flower, arguably the best museum curator last century: all shared Huxley's financial fears and "professionalising" concerns. All wanted Natural History knocked into shape as a professional (i.e. reputable) discipline, which invariably meant ousting the spider-stuffers, country parsons and the like. Christian surgeons, the affable Flower and Huxley's Calvinistic colleague, J. W. Hulke, who did so much of the palaeontological donkey work, admittedly kept a low profile, nevertheless their presence was probably crucial. From this fact alone it should be clear that what we are investigating is not the wearisome 'warfare' history of old. Never was it simply a matter of Church-baiting rationalists triumphing over religious obscurantism, but a more subtle attempt, jointly undertaken by 'agnostics', deists and some Christians, to professionalise science and put it at the disposal of the mercantile middle classes. Hence the inevitable opposition of the Oxbridge dons, in the service, so to speak, of the traditionally antagonistic landed party, with its diametric social and moral order. Not to put too fine a point on it, the debate – even as it affected palaeontology – was probably as much about the kind of social order that should prevail, democratic or aristocratic, as whether natural theology still had a place in evolutionary theory, although in the final analysis the two factors will undoubtedly be found to be inextricably related.

So my strategy, broadly speaking, will be to investigate how far abstruse debates over mammal ancestry or dinosaur stance reflected the cultural context and the social commitment of the protagonists, and as a result to determine the extent to which ideological in-fluences penetrated palaeontology to shape it at both the conceptual and factual level.

Huxley, Owen, and the Archetype

T. H. Huxley was described by Darwin as his "devil's disciple" and Richard Owen once rated "England's greatest comparative anatomist". The two men were acknowledged world leaders in palaeontology; they worked within a mile of one another for the best part of thirty years, and yet, the physiologist Michael Foster admitted in 1869, had they tried to co-author a *Principles of Biology* they would never have got past the title.[1] Their irreconcilable differences are the subject of this chapter – differences, mistrust, and finally hatred which left neither man able to speak to the other for three decades. This would be relatively uninteresting, however, if it did not form the basis of a deeper study of the way social and cultural factors can shape science itself. So we will use the growing acrimony as an *entrée* into the social construction of palaeontology, like a wedge, as it were, to open up the wider aspects of its conceptual foundation.

In this respect, we can expect some help from the sociologists of knowledge, in particular, from the thesis of Barry Barnes, Steven Shapin, and the Edinburgh School, that ideas can be used as tools by social factions to further their own interests. (Only in our case, the 'ideas' are likely to be in more concrete form, as in the case of fossil reconstructions.) We will also take a fresh look at the problem of personalities, while trying to avoid the obvious pitfalls of the psychological approach. Too often, wider social concerns have been obscured by historians preoccupied with reslaying the slain, who drag personalities to the fore and use character assassination in place of a fuller contextual understanding. Perhaps this is excusable in Owen's case; it is not easy to love a man who recalled Cuvier's prophecy of a future Newton of Natural History and practically offered himself for the post.[2] Nothing was more calculated to instil angry misgivings in contemporaries, and in none more than Huxley, who recognised Owen as "the superior of most" but was galled by his refusal to let anyone forget it. Since it was Huxley's circle which rewrote much of Victorian science history, we have the daunting task of sweeping aside the partisan myths. Owen lived to be 88, was still publishing in the 1880s and died only in 1892. In the later years he was treated as the Grand Old Man of British biology, a sort of palaeontological Gladstone. In fact he died just three years before Huxley, and our job is not made any easier by Owen's Reverend grandson asking Huxley to round off *The Life of Richard Owen* with an account of grandfather's character and achievements (to make it "as impartial & authentic as possible", he explained[3]). Huxley was astonished and politely refused the first part on the grounds that "I was never in your grandfather's house nor he in mine: and I doubt if we met in private society more than half a dozen times". On the second, Owen's "scientific labours", he did contribute a chapter, straining to be fair but letting the strain show, and fully expecting "poor old Richard's ghost" to haunt him for his efforts.

Our first task will be to examine the problems facing scientists in the crucial years, 1850–1859, i.e. prior to the publication of Darwin's *Origin*, which should give us the material to tackle the development of Owen's palaeontology in the next chapter. We will then be in a position to discuss the wider social issues and transcend the naive view that Owen's dissent from Darwinism stemmed from bigotry or obscurantism.

The task is formidable. An inordinate amount of sympathy is needed in dealing with Owen. His thought might represent "a milestone in the history of biology",[4] as Roy MacLeod says, but with the man "both feared and hated" his contemporaries often found themselves unable to separate their feelings for him from their judgment of his science. Clearly the time is long past for a thorough re-examination; we need to reconsider the dogmas of the older generation of partisans, those who seized on Huxley's indictment to build a damning case. Owen was "a social experimenter with a penchant for sadism and mystification" in William Irvine's

Apes, Angels, and Victorians; his flagrant professional misconduct "dictated by an arrogance and jealousy" according to Darwin's biographer Sir Gavin de Beer. While not necessarily doubting these subjective statements, one yearns for a more productive way of treating the inflamed situation. And one which, to be fair, allows us to investigate the action of vested interests, not only in Owen's case, but Huxley's also. Both, after all, were out to build a socially-useful science, even if each served a different section of society. So Huxley's fears, compensating strategies, and their social repercussion must bear equally close scrutiny.

This is especially necessary as in the past Huxley has been pictured as ideologically untainted, or nearly so – on the 'right', i.e. evolutionary, path and therefore guided by the highest motives, the search for truth. We learn that he was blocked or hounded at every turn; and with Owen notoriously pictured as at best untrustworthy and at worst treacherous, morally indignant evolutionist-historians have had a field day denouncing him as an "aggressor". Leonard Huxley used chilling prose to describe his father "dealing a blow which would weaken the authority of the aggressor",[5] and in the Cold War years when Irvine and de Beer were writing this became the fashionable attitude. But it lacked a certain sophistication and did little to further our understanding of the great under-currents of Victorian thought, or of the industrial and economic movements which were changing the face of science.

Manuscript sources at our disposal (Owen's twenty-seven volumes of letters are housed at the Natural History Museum in South Kensington, and Huxley's just along the road at Imperial College), make it relatively easy to dismiss the wilder claims – that an obstreperous Owen tried a number of dirty tricks to forestall Huxley (de Beer) or prevent him publishing (Huxley's biographer Cyril Bibby). Oddly, after so much time, feelings still run high against Owen. Recently Michael Ruse in *The Darwinian Revolution* (1979) had the temerity to turn historiography on its head and suggest that from the first Huxley had "hounded the older man".[6] Not to be rudely inverted, Michael Ghiselin in a review vigorously restated the traditional case: "The fact remains that Darwin's bulldog was kicked when he was a puppy. One would only expect him to fang the malefactor whenever the opportunity presented itself." But unpublished sources throw considerable doubt on whether such neat and simple solutions will ever be found. Personal animosity there was, but merely to determine who struck the first blow is surely futile, without at least asking *why*, which should expose the mitigating circumstances.

My strategy for the moment is to concentrate on the financial and professional worries suffered by a good many scientists in the fifties, and the frustrations these caused. This will allow us to connect in later sections with one of sociology's current concerns: the incipient professionalisation of science in the mid-Victorian period. What does emerge from manuscript evidence is a picture of personal, professional and scientific frustrations of immense complexity and demanding more subtle treatment than has hitherto been accorded.

The Roots of Antagonism

> He is not referable to any "Archetype" of
> the human mind with which I am acquainted.
> > Huxley on Owen's mentality,
> > in a letter to Edward Forbes,
> > 27 November 1852.[7]

Richard Owen (1804–1892) was physically distinguished: a "tall man with great glittering eyes", Thomas Carlyle called him, which was a polite way of putting it, since young Huxley saw him as a "queer fish, more odd in appearance than ever".[8] A product of Lancaster Grammar and the Medical Schools of Edinburgh University and St Bartholomew's Hospital, Owen eventually settled in the "happy Fields" of Lincoln's Inn, at the Royal College of Surgeons, marrying the daughter of William Clift, the Conservator in the Hunterian Museum. He was nothing if not industrious, and living quarters above the Museum enabled him to work through the night when needs arose. He published prolifically throughout the thirties and forties: on the deceased inmates of the Zoological Gardens, the reproductive organs of marsupials and monotremes from the Australian colonies, on the great apes, dinosaurs, and the extinct flightless birds of New Zealand. Yet throughout his work there was a leit-motif: in the radical thirties Owen, like so many contemporaries, wrote specifically to repudiate the heresy of transmutation, structuring his ideas on apes, dinosaurs and the platypus accordingly (a subject we will return to). By the mid-fifties, he had been two decades as Hunterian Professor, with "a European reputation, second only to that of Cuvier".[9] Flower saw him at his "zenith" in 1856 and commanding considerable respect from continentals. Yet the financial rewards were few: a Hunterian salary of £300 p.a., which in the words of historian Geoffrey Best "did not carry a family man far up the slopes of gentility", topped up by a Civil List pension of £200 arranged by Sir Robert Peel. This might have been better than average for English science, but it still only amounted to what an Assistant Clerk in the civil service could expect to earn. And in other ways Owen was luckier than most, having been presented, for example, with Sheen Lodge in Richmond Park by Queen Victoria in 1852.

For all that, Owen was uncomfortable at the College during the fifties, and desperate to transfer to the Natural History Department of the British Museum. The exact reasons are in doubt; according to Huxley, Clift had freed Owen from the drudgery of administration, giving him more time to research unhampered, "but after Clift's retirement, things began to get in a dreadful muddle which caused the disagreements with the Council of the College that led Owen to move heaven and earth to get the appointment made at the British Museum."[10] A vacancy had arisen at the museum in 1851, and some even wondered "whether the whole establishment had better not, *quoad* Zoology, be remodelled and placed under Owen's superintendence." Something similar was in fact to happen, but not for another five years. Huxley, fresh from his four-year survey expedition to New Guinea, was astounded at the "heartburnings and jealousies about this matter". Evidently those ensconced in the museum feared an erosion of their power and it was thought that uproar would ensue if Owen got the job. Instead the post fell to George Waterhouse, who had the backing of the influential geologist Charles Lyell; and Owen, ostensibly for financial reasons (it paid less), withdrew and disclaimed all interest, although apparently remaining on the best of terms with Waterhouse. This was Huxley's first taste of corridor-politics and it both intrigued and appalled him, his letters to W. S. Macleay in Sydney and his sister in Tennessee becoming full of hearsay and horror stories about such machinations.

The problem is to *evaluate* Huxley's judgments rather than repeat his words verbatim, and this means setting them against a framework of his own social and scientific preconceptions. Concentrating on the year 1851, for instance, we find that the twenty-six year-old was jobless and almost penniless as he watched Owen angle for the plush Bloomsbury position, a circumstance which might explain his lack of sympathy. In the same year, Owen had published a damaging critique of Lyell's science in the *Quarterly Review*, regretting that a man in Lyell's position should have sunk to special pleading on behalf of a redundant steady-state geology, and refuting his claim that the fossil record showed no progressive trends.[11] Lyell was furious, probably half-believing Mantell, who told him that Owen's "spite" was retaliation for having backed Waterhouse for the British Museum. But we must remember that Owen spoke for many geologists in 1851 (Darwin included), whereas Huxley was that rare breed, a Lyellian non-progressionist (Chapter 3), which might help explain why he took the review amiss. So we should beware of criticising Owen solely on Huxley's say so. Finally, Owen had become embroiled in a bitter priority dispute with Mantell that year over who was to describe *Telerpeton*, a fossil reptile disinterred from the Elgin sandstones of Scotland. Owen rushed into print leaving Mantell seething; one lampoon went so far as to suggest that the affair "worrited him to death" (actually, a long-standing spinal injury had left him "almost frantic with pain" and destined for a melancholy life sniffing chloroform before dying in 1852). This of course is the reason why he attributed base motives to Owen in the Waterhouse incident, and why Huxley should single Mantell out as Owen's "arch-hater". However, the rights and wrongs of the case are far from clear cut. Indeed, Michael Benton, who specialises in Elgin reptiles, has recently reinterpreted it as a bureaucratic mix-up after discovering that in fact both Owen *and* Mantell had been asked to notice the fossil,[12] all of which suggests that they were really the victims of circumstance and their own obsessive natures. But to my way of thinking what ruffled most feathers was not Owen's alleged unethical conduct so much as his callous indifference. With Mantell barely cold in his grave, the *Literary Gazette* carried an atrocious obituary, thought to have been by Owen, which desecrated the poor man's remains and churlishly pointed up Mantell's "overweening estimate" of his own importance – a tasteless piece which in all probability lost Owen the Presidency of the Geological Society that year.[13]

Such "literary pettifogging" was the way the scientific world evidently looked to Huxley in the early 1850s: the corridors of power whispering with intrigue and jealousy, necessitating those "little stratagems" which at once horrified and delighted him. "And yet withal you will smile at my perversity", he told his sister, "I have a certain pleasure in overcoming these obstacles, and fighting these folks with their own weapons"[14] – an admission which tells us a good deal. Not only was he intimately aware of the politicking in high places, but was willing to play the game.

Thomas Henry Huxley (1825–1895) was twenty-one years Owen's junior, a fact one is apt to forget. It meant that he missed the conservative scientific reaction during the trouble-torn thirties, still being at school in Ealing, while from 1846–1850 – when Owen attempted to romanticise British biology against a backdrop of Chartist agitation – the young Huxley was an Assistant Surgeon RN on equatorial shores, or investigating drifting marine life from the decks of HMS *Rattlesnake*. For four years after his return he faced the uphill struggle to become established, and was driven to almost schizoid extremes. At times he was brash and self-assured (what other Assistant Surgeon would go straight to the First Sea Lord and demand money?), at others racked with doubt and despondent at the lack of pay or prospects.

Huxley's attempt to become established provides a fascinating case study. He was in a precarious position, torn between social subservience and professional integrity: dependent on senior scientists, who held the purse strings and whose voice, as Forbes said of Owen, was "of importance in all Government matters", while simultaneously and often tactlessly engaged in demolishing what he considered the tattered superstructure and antiquated metaphysics of 'official' science. Look first at the way Huxley tugged at the purse strings. He needed two or three hundred pounds to work up his *Rattlesnake* researches, but was having excruciating difficulty in obtaining it. The Royal Society did receive a Government grant of £1000 a year but turned him down in 1851 because he was still in the Service, which made it the Admiralty's duty to finance publication. So he switched targets and for two fruitless years harangued their Lordships. Then, about 6 November 1852 he asked Owen for another testimonial, this time to send direct to the Secretary of State

(Walpole). After ten days there was no sign of it, and an irritable Huxley was consoled by Forbes: "Ask Owen again. I do not think the forget is intentional & am very sorry to hear of it because I regard it as an unpleasant symptom of the physiological condition of Owen's memory."[15] So Huxley dropped Owen a reminder, and on the 27th reported to Forbes on the upshot:

> I wrote to him as you advised and got no answer whatever. Of course I was in a considerable rage. I then on the fourth day afterwards, met him. I was going to walk past, but he stopped me, and in the blandest and most gracious manner said, 'I have received your note. I shall *grant* it.' The phrase and the implied condescension were quite 'touching' – so much that if I stopped a moment longer I must knock him into the gutter. I therefore bowed and walked off. This was last Saturday. Nothing came on Monday or on Tuesday, but on Wednesday morning I received "with Prof. Owen's best wishes", the *strongest and kindest testimonial any man could possibly wish for*! I could not have dictated a better. I immediately sent a copy to Mr Walpole.
>
> Now is not this a most incomprehensible proceeding?
>
> I gave up any attempt to comprehend him from this time forth.

Forbes reluctantly agreed and puzzled over Owen's psychology:

> He is certainly one of the oddest beings I ever came across & seems as if he was constantly attended by two spiritual policemen, the one from the upper regions & the other from the lower – the one pulling him towards good impulses & the other towards evil. As I believe men's bad qualities in 3 out of 5 instances are generated in the stomach I lay many of O—s various eccentricities to the charge of his ill health. He has very much that is good & kind in him, with all his faults.

One can appreciate Huxley's frustration and rage; for two years he had tried every avenue to obtain his money. Cajoling Edward Sabine of the Royal Society and the First Sea Lord had proved equally futile, even with the backing of Owen and Forbes. Looked at from Owen's side, however, things take a different hue. This testimonial was one of a string he had unstintingly provided the young anatomist. It was largely Owen's doing – he appealed direct to the Lords Commissioners – that Huxley was transferred to the Guard Ship *Fisguard* berthed at Woolwich in the first place[16], which effectively allowed him leave of absence on half-pay to write up his researches.

Huxley's frustration was compounded by his failure to obtain any teaching post throughout the period 1850–1854. At each attempt, Owen backed him to the hilt. Huxley tried for Toronto in late 1851, which was paying the Hunterian-equivalent of £300 p.a., and Owen testified in glowing terms to his "high learning" and "skill as an original observer". Again, on 30 October 1852 he praised the "philosophical spirit" of Huxley's memoirs in an abortive effort to push him into the Aberdeen chair.[17] Note the date – exactly a week before that request for a grant testimonial which Owen failed to supply with the necessary haste. Well might he have been getting a little lax, with requests arriving a week apart. Supplying the young anatomist with references had become routine; he had been doing it for two years, to no avail. For his part, Huxley was getting desperate, and his utter inability either to obtain money from the Exchequer or an academic post left him bitter and in two minds whether to emigrate. His finances had remained precarious since arriving home from New Guinea a hundred pounds in debt. It had been two years since he had seen his fiancée, and he badly needed a paying position in order to bring her over from Australia. "Sometimes I am half mad with the notion of burying all my powers in a mere struggle for a livelihood", he wrote to her that August. "Sometimes I am equally wild at thinking of the long weary while that has passed since we met."[18] What James Paradis in his important study *T. H. Huxley: Man's Place in Nature* (1978) identifies as Huxley's "chronic mental depressions and collapses, his sense of isolation and trauma", was already in evidence. He began to doubt his scientific vocation, and once even talked of "having failed in the whole purpose of my existence". In April 1852 he weathered the shock of his mother's sudden death, only to watch his father become "completely imbecile" and lose any feeling "beyond a vegetative existence". He reached a state of mental exhaustion.

Despite official recognition – an FRS before the age of 26, praise for his papers, and the accolade of a Royal Medal in 1852 – he remained desperately poor. "I can get honour in Science, but it doesn't pay, and 'honour heals no wounds'." He was naturally sensitive to Owen's condescending attitude, growing resentful and in the end suspicious. Indeed, the continuing indictment of Owen, "infamous to the backstairs of science", partly follows from Huxley's accusing finger. He imagined this "particular friend" pooh-poohing one paper "to a dead certainty" should it fall his way[19] (not in itself paranoic: with Huxley tearing the heart out of Owen's prized theses, he might have expected hefty opposition). Nonetheless, from such stuff are myths made. What was simply a fear became something more in the hands of historians. Bibby imagined that "the increasingly jealous Owen" *had actually* "tried to prevent publication", a spurious leap which reinforces the whole wretched stereotype. At the same time Huxley envisaged Owen putting up 'No poachers' notices around his preserve, barring admission to aspiring anatomists. And thoroughly frustrated in his own efforts to get a Royal Society grant, he naturally wondered where Owen was obtaining *his* funds. Huxley tackled Sabine again in October 1853, declared himself "out of all patience", and evidently enquired into Owen's arrangement, at least for printing his paper on the Ground Sloth recently received from Buenos Aires. Sabine stopped him short:

> You are not rightly informed about Owen's paper on the Megatherium. No money has been allotted to that from the Govt. fund of £1000. The last I heard of it was that the Council of the R[1] Society agreed to what they understood to be Mr Owen's desires, viz that it should be divided into 3 or 4 papers, to come out in successive numbers of the Phil Trans (successive years), the cost in each year to be £50 or so, which they however considered the extent to which they would be justified to go in the case of a single paper. But Mr Owen's reply was that the publication would require £700 at the least, whereupon it was dropped.[20]

So some of Huxley's suspicions at least were unfounded. Owen was not appropriating the lion's share of the funds; in fact, he was having as much trouble meeting publication costs as the next man. In the end Owen *was* forced to break the paper and spread it over six years. And long before the end Huxley had himself been paid. He was struck off the Navy List in March 1854 for refusing to return to his ship (in protest at not being paid). No longer in the Service, he was promptly funded by the Royal Society to the tune of £300, and that July finally found the security he needed in Forbes' old post at the School of Mines.

None of this is brought forward to whitewash Owen – his known quirks would make any such endeavour self-defeating – only to point out that Huxley's innuendo often served to incriminate him when the exigencies of the situation were to blame. Owen was unlikely to have "pooh-poohed" the molluscan paper, but since he did not referee it, we will never really know. We *do* know that he received his complimentary copy with "great pleasure" and rejoiced at "this additional evidence of the scientific activity of one who works so much after my own heart".[21] Of course this may have been routine politeness, civility in exchange for a free paper; it is harder to see as a snide disguise of his real feelings, and I am not sure that we need to turn ourselves inside out to discover some hidden meaning. Owen does seem to have been genuinely supportive, initially unaware of the damage being done as Huxley began chipping away at his prized theories.

So what of Huxley's science, to what extent was it a needling point? Perhaps the best way into the subject is to tackle Owen's views on species origination and particularly his response to the much-abused *Vestiges of the Natural History of Creation* (1844).

How were New Species Introduced?

. . . like the "reputed Author of the 'Vestiges'" I also feel very eager and curious to know the several and more probable modes whereby successive species have been introduced on the planets, and sincerely hope you will resolve ere long to make the public acquainted with your knowledge and views on this interesting subject.

John Chapman to Owen, 13 January 1848.[22]

That "beastly book" (as the Cambridge geologist Adam Sedgwick called the anonymous *Vestiges*) was written by the Edinburgh publisher Robert Chambers to prove that creation by natural law was more befitting of Omnipotence. The Almighty had staggered creation by a piece of master programming, employing a "higher generative law" to suspend normal reproduction and trigger the birth of more advanced species at the appropriate time. It was this, the insensible blending of discrete concepts, "law" and "development", that so incensed Sedgwick (he considered it "the most unspeakably preposterous instance of bad reasoning in the whole volume"). His blood boiled at the thought of fashionable Londoners falling for such philosophic nonsense, and in an overkill review he stamped with "an iron heel upon the head of the filthy abortion, and put an end to its crawlings".[23]

Historians have invariably accepted this as the official line towards the *Vestiges*, which in some ways it was. But the book was a massive best seller: four editions were snapped up in seven months, and by 1846 two editions of the explanatory *Sequel*. And not all of it was popular appeal; there were compelling reasons why it was welcomed by some of the London intelligentsia. Radicals, late romantics, and disillusioned Tractarians discussed and sometimes praised the book, less for its scientific insights than its unorthodox approach. Even Huxley's close friend Forbes, while deploring the half-cocked blunders, thought it came "like a breath of fresh air to the workmen in a crowded factory".[24] It was a talking point, especially, at the weekly *soirées* in the Strand organised by John Chapman, described by Spencer as the "only respectable publisher" of books "tacitly or avowedly rationalistic" (he published Spencer's *Social Statics*). Later, in 1851, Chapman was to buy the rationalist *Westminster Review*, a journal which had already treated both *Vestiges* and its *Sequel* kindly, and in 1853 he persuaded Huxley to undertake a regular "Scientific Section". It will pay us to concentrate on Chapman's friends, first because Sedgwick's unmitigated scorn, coming from a Cambridge don, would have found little favour among them, and second because they knew and questioned Owen on the *Vestiges*, and his position has often been seen as enigmatic.

By 1849 Spencer was attending the *soirées*, his offices at *The Economist* being just over the road. Here he was introduced to members of Chapman's circle: the lapsed Anglo-Catholic J. A. Froude, Francis Newman (brother of the Cardinal), the writer and positivist G. H. Lewes, and the novelist George Eliot, who lived for a time in Chapman's house, befriended Spencer, and ended by living with Lewes. Even Owen, as a writer, occasionally found his interests best served by siding with this group. Serious publishing was a liability and Owen admitted that his first concern on starting a new book was whether he could afford the loss. So he supported Chapman and Spencer in their campaign to abolish price-fixing, in the hope that an open market would encourage sales.

A good deal of dissolvent literature stemmed from this group. Froude in *Nemesis of Faith* (published by Chapman himself) and Newman in *Phases of Faith* spearheaded the ethical revolt against orthodox Christianity. They rejected doctrines like the Atonement and Damnation which grated with the meliorist climate of the age (the faith that man's lot can be improved in this life without waiting for other-worldly salvation). No wonder there was considerable interest in Chambers' optimistic picture of progressive development, with its onward and upward sweep towards a higher existence. And at least one member of the future *x* Club, the mathematician Thomas Archer Hirst, was suitably impressed by Chambers' solution to the time-honoured problem of evil: i.e. that it resulted from the 'accidental' intersection of laws unleashed at Creation, and was somehow extraneous to the original plan (Darwin was to employ an identical argument as late as 1860).[25] So ethics for some was a major drawback to Christianity. Darwin himself rebelled at the thought of his father and brother in Hell; and the indescribable horror of the ichneumon grub eating its host caterpillar alive caused him more loss of faith than any amount of evolution. Ethics could equally dictate one's choice of science: Owen, for example, rather than lose his faith, explicitly adopted the automaton theory of lower life in his 1842 Hunterian lectures, to save conscious polyps from the "shrinking terror" of being grazed alive by predatory fish.[26] But when it came to the *Vestiges*, the ethical appeal of law for Owen was vastly outweighed by the morally-degrading mechanism of transmutation. He might have embraced law, but the prospect of human brutalisation was anathema; he needed a rival, natural and non-bestial mechanism. Naturally Chapman and his friends looked to Owen for an answer to this great question. Spencer sat in on Owen's lectures in 1852 hoping to find some proof of development, while Chapman begged to be allowed to quote Owen's view as "yours is so very important as bearing on the question." But this was one of a string of pleas Owen received, either to repudiate the "beastly book" (from orthodox scientists: Sedgwick, Murchison,

Sir Philip Egerton, Whewell), or to state his alternative. And always he procrastinated, totally confounding Sedgwick and Whewell by refusing "to appear as having directly or personally aided in any thing that may be regarded as a refutation or antidote to 'Vestiges'."[27]

His reticence has engendered a number of conflicting claims. The conventional line is toed in Owen's eminently Victorian *Life*, where the Reverend Owen concluded that grandfather "had a certain leaning" towards Chambers' theory but declined to join issue for lack of proof. And Owen did indeed send the anonymous author an almost congratulatory letter, admitting that he perused the book with "pleasure and profit", adding that

> no true searcher after truth can have a prejudiced dislike to conclusions based upon adequate evidence, and the discovery of the general secondary causes concerned in the production of organised beings upon this planet would not only be received with pleasure, but is probably the chief end which the best anatomists and physiologists have in view.

Owen did continue with a list of criticisms, but in such diplomatic tones that Chambers could practically have counted him a convert. In 1977 John Hedley Brooke challenged this easy interpretation, and stood tradition on its head by suggesting that in fact Owen found the book beneath contempt and "too mischievous to publicize further by hurling missiles in print".[28] Brooke sees subtle sneers even in Owen's apparent pleasantries, and supports his case with two letters in which Owen told Whewell that nothing could be more calculated to "add greatly to its mischief" than the dignity of a reply. Leave well enough alone was the warning, for fear of calling forth still more editions to answer the charges – a sound caution, for the *Sequel* answered Sedgwick point by point and began shoring up the theory to such an extent that, said one critic in 1846, the "deformities no longer appear so disgusting".[29]

Doubtless Owen wanted the Vestigian infant abandoned in its cradle to die quietly, without the public inquest that would attend a strangulation. But need we really see nothing but ridicule in Owen's response? Treating it as a blanket condemnation, we risk riding roughshod over the finer issues. I will suggest (Chapter 4) that Owen had been deeply worried for over a decade by his social and professional rival, the Professor of Zoology at University College, Robert E. Grant, and I believe much of Owen's science can be understood as a reaction to Grant's atheistical leanings. From this episode, we know the sort of preconceptions Owen brought with him into the Vestigian period. He had an almost pathological hatred of 'Grantian' transmutation, with its dread self-sufficiency and ungodly aspect. Anti-transmutation had been a perennial theme in Owen's lectures, highlighting his repugnance of mankind's supposed ape ancestry. Moreover Owen had demolished the heretical premises seven years previously, in the 1837 Hunterian lectures, when he showed the illegitimacy of stretching the "organizing energy" from one individual to the next.[30] To imagine species 'growing up' to human maturity was metaphoric nonsense. And in 1835 he had gone to pains to dissociate apes and men on anatomical grounds.

Examination of Owen's letters on the *Vestiges* question again shows that his primary target was *transmutation*, particularly in its application to man. He told Whewell that no refutation would convince someone who believed "that his great-great-great-etc. grand father was a Baboon, and that his great-great-etc. grand mother was a Chimpanzee", while replying to the author of *Vestiges* he denied the "origination of man" from apes and repudiated the supposed similarities. Perhaps the tone was less harsh because he was no longer dealing with Grant's godless biology. The Gower Street professor had seen life steered by 'random' climatic changes; not only did the system move under its own impetus, but accidents directed life's course and man was basically a chance outcome.[31] The *Vestiges* was in utter contrast: Chambers pictured life's progress as an unfolding of a Divine plan, designed to place man on the earth as "the unmistakable head of animate creation." The book was primarily a protracted argument for uniformity, and since Chambers conceived law as a Divine mandate, universal progress was a manifestation of Supreme Intelligence. Owen might have lamented the mistakes, hated the transmutation, and even doubted aspects of Chambers' theodicy; yet he was in total agreement on the need for uniformity, and on its correct Providential interpretation.

The *Vestiges* must have struck Owen as a hotch-potch of philosophical issues, some – uniformity, Providentialism, lawfulness – perfectly respectable in the right hands. So my guess at what he was telling the anonymous author is this: 'The best naturalists seek to understand nature through secondary causes – and like you they interpret them as manifestations of Divine Will. You unfortunately confuse the real causes with transmutation, and this is unforgivable'. But this would have left a monumental question mark: 'What are the real causes?' Chapman wrote begging Owen to speak out on "the several and more probable modes whereby successive species have been introduced on the planets" – a reference to Owen's claim that, hard pressed, he could come up with *half-a-dozen* "possible secondary causes", and that transmutation, "which is always coupled with the idea of a specific direction – viz. *upwards*", was "the least probable of the six".

> When I remarked to the (reputed) author of the "Vestiges," the last time he visited the museum, how servilely the old idea had been followed . . . – viz. of "progressive development" – and that there were five more likely ways of introducing a new species, he asked suddenly and eagerly, "What are they?" I declined to give him the information, but shortly after brought prominently under his notice the facts that might have suggested one, at least, of the more likely ways. He saw nothing of their bearing, and I shall refrain from publishing my ideas on this matter till I get more evidence.[32]

From *today's* perspective Owen's admission sounds astonishing. Of course, it might have been a bluff, an attempt at self-aggrandisement, something which fits his reputation for belittlement; or a warning that dabbling in unfounded secondary causes was an exercise in futility. But suppose for a moment that he was telling the truth; suppose that he had located even *one* plausible alternative. If Owen was sincere on greeting the *Origin*, when he conceded that Darwin had tackled

> . . . the question of questions in Zoology; the supreme problem which the most untiring of our original labourers, the clearest zoological thinkers, and the most successful generalisers, have never lost sight of . . .[33]

and if Owen was here referring to himself, then he had never lost sight of the "supreme" problem. Moreover it does us no credit to take the Whig option and insist that he was chasing a will-o'-the-wisp, that any alternative must have been illegitimate because evolution turned out to be 'true'. It is a poor view of history which scores the past on the criteria of the present.

So what *was* Owen thinking of in 1848? Escorting the "reputed author of the *Vestiges*" round the Hunterian Museum, he had evidently stopped to point out a sequence of specimens which should have made him think twice. In his papers, Owen throws out strong clues as to what these were – he even resorts to them in his review of the *Origin*, in a vain attempt to convert Darwin. Also we know the constraints Owen envisaged for any explanation of species: it had to conform to known laws of generation, and as such be seen in operation today. The key, Owen guessed, lay in a subject of great contemporary interest – Alternation of Generations.

Owen had become fascinated in this following the discovery, in about 1842, that parasitic flukes were not spontaneously generated in the gut, but part of a complicated lifecycle involving two, three or more hosts. Very simply, the eggs hatch into free-swimming 'infusoria', which locate and bore into a water snail. Here they develop into larvae, each of which buds off innumerable daughter cysts. And within each of these grows a number of tailed 'embryos', which bore out of the snail. These cercariae, thousands from each initial egg, climb grass stems to be ingested by grazing sheep (in the case of the liver fluke), and here in the final host they become the familiar gut parasite. Thus the cycle is completed, perhaps years later and after untold 'generations'. It was not so much a metamorphosis, which described an individual's change, e.g. from tadpole to frog. Owen called it a *metagenesis*, since numerous generations were involved. He thought the cycle "astounding" for the profound differences between the generations, which in the past had not only been classified as different organisms, but sometimes placed genera or orders apart. He realised that the cycle from 'infusorian' to snail parasite, tailed cercaria, and fluke gave the appearance of breeding successively higher types. *The process simulated transmutation* – but without the gradualness. He even admitted that "A partial knowledge of the strange phenomena of metagenesis might at first be *mistaken* for direct evidence of 'transmutation of species'", and yet the progressive 'births' obeyed ordinary laws of generation.[34]

Owen's study of flukes and medusae led to his book *On Parthenogenesis* in 1849. Historians often side-step this work because it apparently lacked any relevance to higher issues. But Owen attached great weight to it. He told Huxley it was "one of the best things he had done", although Huxley failed to see its underlying worth, and to the end of his life denied that Owen had added anything "to the common stock of observed facts" on the subject. But suppose Owen partly valued the book for another reason, for its potential solution to life's ultimate mystery. He admitted to the British Association in 1858 that

> The first acquaintance with these marvels excited the hope that we are about to penetrate the mystery of the origin of different species of animals; but as far as observation has yet extended, the cycle of changes is definitely closed.[35]

The fluke cycle always returned to its start, however many years later, and Owen only mooted breaking the chain in 1860, upset by Darwin's denial of any alternative to transmutation except a miraculous flashing together of atoms. Under peculiar conditions, the individuals might break free and go on breeding their own kind, in which case four or five new genera, or even orders, might emerge. Darwin missed the point, as did Owen's guest at the Hunterian fifteen years earlier (so far as I can see only the German idealist von Kölliker took up the question). Owen himself admitted that it was only a speculation, though "on a foundation at least as broad" as Darwin's. But since the generations alternated before our eyes, it did give naturalists something to study, whereas Darwin left the science of origins "very nearly where [he] found it."

Treated sympathetically, Owen comes across as a psychologically complex character struggling with immensities – driven by his hatred of ape ancestry to search for a sensible non-transmutational answer to the biological challenge of the age. Unfortunately, hindsight gives us a vested interest in championing the transmutationists of history, yet Darwin's contemporaries were casting their nets wider in search of a solution. Metagenesis suggests that a rival was at least *conceivable*, and should caution us against reproaching early Victorians for an apparent crisis of nerve. No longer can we agree with Neal Gillespie in *Charles Darwin and the Problem of Creation* (1979) that "A truly natural cause of new species could only mean, as Darwin realized, descent, and that could only mean transmutation . . ." Nor necessarily with his harsh judgment that the "equivocation" of scientists was "an evasion of the problem rather than a strong determination to work out a natural solution of it."[36] Owen showed, if nothing else, that natural causes need not mean transmutation. And considering the overpowering objections to at least the Grantian mechanism, with its spontaneous generation and accidental course, "need not" should read "emphatically did not".

Huxley's Early Science

These points are brought forward as a caution against deprecating the upholders of law. But they also show that with the metagenetic edifice having tantalising possibilities, Owen might have been doubly irritated by any tampering with the foundations. Yet by 1851 Huxley's demolition work had begun. Having made a point of studying the perishable polyps and medusae on the *Rattlesnake* voyage, he had returned home convinced that the theory of alternating generations was fundamentally misconceived. He also returned with Macleay's note of introduction to Owen, requesting that he help the Assistant Surgeon publish, since his researches bore so intimately on "the subject of your 'Parthenogenesis'".[37] This Owen did, only to see Huxley try to demolish the theory. What Owen had taken as discrete individuals, Huxley considered nothing more than wandering sexual organs. According to him, all the so-called generations taken together – that is, all the members of a cycle – amounted to only one individual. These free-living organismic 'fractions' he termed "zoöids", a name Darwin doubted would catch on for "creatures bearing so plainly the stamp of individuality". In fact, apart from his friends George Allman in Dublin and W. B. Carpenter, Huxley seems to have stood alone, and admitted as much in his first public lecture, on "Animal Individuality", delivered to the Royal Institution. Nonetheless, he repeated in 1851 and 1852 that the "whole theory" of alternating generations "must necessarily fall to the ground", and at the Royal Institution singled out Owen's book for failing to meet the facts.

According to the prevailing caricature, Owen refused to let anyone, least of all a mere tyro, steal his thunder, and he was not above back-stabbing if needs arose. But in this instance he failed to strike back. In January 1852 he again offered "to do anything in his power" to help Huxley,[38] and a year later evidently *still* considered him a man after his own heart. Only after some facetious remarks by Huxley a couple of years later did relations lose all semblance of cordiality, suggesting that it was Huxley's lack of civility that Owen finally found intolerable. In a slashing review of *Vestiges* in 1854, Huxley hauled Owen over the coals, in what was a slick operation for an inexperienced young reviewer. Although Chambers himself never guessed the authorship (at least, he invited Huxley to stay the following year), Darwin, and I dare say Owen, did. "By Heavens," wrote Darwin, "how the blood must have gushed into the capillaries when a certain great man (whom with all his faults I cannot help liking) read it!"

A situation already tense after Huxley's attempted refutation of 'Alternation of Generations' suddenly deteriorated as the two headstrong men were thrown into a petty world of tit-for-tat. Accusations flew thick and fast, of Huxley's "blindness" and Owen's self-glorification, and Huxley ended up slating Owen's *Comparative Anatomy and Physiology of the Invertebrate Animals* (1855) in a review which, said Carpenter, will show up Owen's "ignorance, prejudice, or ill-nature most conspicuously".[39] Relations had so degenerated that Huxley formally severed them in December 1856 on the grounds of Owen's lapse of protocol: entering himself in the *Medical Directory* as 'Professor' in Huxley's School of Mines, on the strength of being allowed to deliver a dozen lectures on "Fossil Mammals" in the theatre. Huxley considered it a slap in the face and told his friend Frederick Dyster, "Of course I have now done with him, personally. I would as soon acknowledge a man who had attempted to obtain my money on false pretences".[40]

Distilling this personal hatred out of Huxley's dealings with Owen would prove well-nigh impossible. At times it permeated every action, every thought, and Huxley thrived on the tension. He eventually placed Owen with Gladstone and Wilberforce in that demonic trinity which for him summed up the cants and hypocrisies of British society. In the final analysis, Owen was notoriously insensitive, and occasionally callous; and despite Forbes' plea that much "is good & kind in him", Huxley invariably drew out the worst. Even continentals sympathetic to Owen's cause were put off by the man himself. The German zoologist Victor Carus confided to Huxley that for all his "admiration" of Owen, he "blushed" at his effrontery. Carus was keen to introduce Owen's osteological ideas into Germany, but first "I think Mr Owen could be a little more polite towards me"[41] – and the fact that Carus ended up translating Darwin's *Origin of Species* and Huxley's *Man's Place in Nature* shows how Owen could alienate even potential allies.

A bearded T. H. Huxley. Judging by the gorilla on the blackboard this photograph was probably taken in the early 1860s. (By courtesy of the Wellcome Trustees.)

But a fine study of personalities can take us only so far. It leaves us
dangling, unable to connect with the social themes in the industrial
age. Personal animosity might explain the intensity of the debate,
but rarely its chosen ground. For that we must understand the
ideological estrangement already in evidence. In particular, Huxley
had grave reservations about Owen's kind of realism, which would
turn laws into commands, archetypes into thoughts, and place the
lot at the service of a disintegrating natural theology; and it was
Huxley's positivist approach which marks him out in the vanguard
of the new middle-class movement in science. So a deeper study is
necessary to understand why he cast Owen as the whipping boy in
his "scientific young England". We will have to examine the
changing power structure within the scientific community itself,
Huxley's "plebeian" commitment, and what Paradis calls his
attempt to create a "democracy of scientific knowledge".[42] And we
must question the political meaning of his Working Men's lectures
and crusades for technical education, which were evidently
attempts to win working class support for the new movement (a
theme continued in Chapter 5). 1855 saw the first of his Working
Men's lectures in the Jermyn Street theatre, delivered, he said, to the
best audience he ever had. And even at this time he found that by
consciously avoiding "the impertinence of talking *down*" to his
"fellows", he could carry them on the most heretical journeys.[43]

Though from an earlier generation, Owen's background was not
dissimilar to Huxley's. He too lectured to mechanics and artisans
and more fashionable society at the Royal Institution, provincial
institutes, and YMCA. But he was clearly feared as a social experimenter. To make matters worse the Reverend Owen ruinously
edited his letters to leave *The Life of Richard Owen* an interminable
succession of bishops and dukes, skillfully manipulating material to
help grandfather jockey his way into the highest echelons. No
doubt the stereotype contains a good deal of truth: Owen seemed
happiest among royals and curiously insecure among colleagues, to
an extent that infuriated the young guard, who no longer wanted
the support of the aristocracy. But the traditional image runs the
risk of becoming wildly lop-sided. We tend to forget that he refused
Sir Robert Peel's offer of a knighthood in 1845, or that he did tireless
work on the Sewers Commission with the social reformer Edwin
Chadwick. We cannot deny that like Charles Kingsley (who paro-
died the Huxley–Owen *fracas* in *The Water Babies*) Owen gave

"edifying little talks" at the Palace (Ghiselin's words), the kind too easily dismissed in a different day and age.[44] Obviously these lacked the social punch of Huxley's *Lay Sermons*. Even when Owen was presented with a heaven-sent opportunity, like Dickens' offer to write for *Household Words* – a middle-class journal designed to deal with social evils like slums and the shelterless poor – he penned even more innocuous pieces on the private lives of zoo animals. Yet he *could* stir up a theological hornets' nest: just consider his 1863 clarion call to the YMCA, which went so far denouncing denominational opposition to a lawful creation that even Lyell's hair stood on end (Chapter 2). Nonetheless there was a widespread feeling of social betrayal, although one obituarist reminded the late Victorians that even Owen's love of the high and mighty did result "in solid gain to science in the acquisition for the nation . . . of the magnificent building in Cromwell Road", Alfred Waterhouse's imposing Natural History Museum – a testimonial of Owen's friendship with Gladstone.

Younger biologists saw a dark side even in this, and once again we can trace the opposition (including Huxley's) to many of Owen's proposals for the new museum to incipient professionalisation. The young guard were seeking the right to self-determination, which meant setting up codes of conduct, raising standards, and selling themselves to the public. Responsibility to one's peers as much as the facts was explicit in this call for higher standards. The prospect of power concentrated in one man's hands had disturbed Huxley since 1851, when Owen's move to the British Museum was first mooted. Carpenter was equally concerned by Owen's plans for a national museum, telling Huxley:

> I am not at all clear that the headship of a single man reporting direct to the Minister would be the best Government of the concern. Doubtless it works well at Kew; but who can find fault with the Hookers? Of course Owen would be the Autocrat of Zoology & Palaeontology; would it not be desirable to make him feel that he is responsible to a body of scientific men, who are competent to estimate and criticise his proceedings?[45]

Archetypes and Laws

But the root cause of Huxley's opposition was ideological. The young guard's trenchant positivism – their attempt to strip off all metaphysical traces and liberate science from priestcraft, leaving it rational, value-free, and serviceable to its new mercantile masters – actually made a confrontation necessary, from a propagandist point of view, because Owen's idealism was a mainstay of natural theology. Huxley therefore slated his "metaphorical mystifications", deprecated his "scholastic realism" for equating laws with Divine edicts, and deplored the Platonic Archetype as "fundamentally opposed to the spirit of modern science".[46] Yet the ideal archetype, the model or prototype vertebrate, played a crucial role in the emergence of modern palaeontology, and Owen's Romanticism deserves thoughtful reconsideration from a non-partisan angle.

Owen was not alone in looking to Germany for a more sophisticated philosophy. In the year of his *Nature of Limbs* (1849), that other 'Germaniser' George Lewes mockingly complained of the "prevailing mania for German philosophy".[47] Lewes was a positivist who saw the practical value of a schematic archetype and actually considered Owen "the greatest living comparative anatomist", tapping his brains in 1852 for a projected biography of Goethe. And yet one doubts that there was any real "mania". The German influence was never strong in England, indeed its attempted introduction was often greeted with hostility. Partly, I suppose, this had to do with the lack of bilinguals: Huxley and Owen were among the few scientists – "the *very* few I am sorry to say", Huxley apologised to von Kölliker[48] – with a gift for the tongue. But Owen's problem was compounded by his coming to *Naturphilosophie* late – the dryly anatomical *Archetype* book first appeared in 1846 and *On the Nature of Limbs* in 1849. By this time *Naturphilosophie* was a spent force in Germany and the scientific materialists already gaining the upper hand. In England by the later fifties, when Huxley and Tyndall first flexed their muscles, Owen's science was already looking tame. He ended his *Palaeontology* in 1860 with a reaffirmation that the "great First Cause . . . is certainly not mechanical". To which a disappointed *Athenaeum*, desperate for some stronger antidote to the Darwinian intoxicants, threw up its arms, "And is this all? Is this the sole residuum of the highest generalizations of your vaunted science"?[49] But the fact that the metaphysical high-spot was greeted with little more than a yawn after Darwin belies its immense usefulness before 1859.

Romanticism was a reaction to the eighteenth century clockwork cosmos and, in early nineteenth century England, to the horrors of industrialisation and its utilitarian supporters. The influential essayist and historian Thomas Carlyle, for example, sought "deliverance from the fatal incubus of Scotch or French philosophy, with its mechanisms and Atheisms", and turned instead to German mysticism. Rosemary Ashton in *The German Idea* (1980) pictures him preaching against "the evils of industrial and materialist society".[50] But to sum up Owen so neatly would be to miss some of the subtleties, even if he was sensitive to the materialist threat posed by his academic rival in Gower Street, the Scots transmutationist Robert Grant (Chapter 4).

By mid-nineteenth century a new factor had given Owen's romanticism a distinctive flavour. Liberalism and political democracy were then making a significant impact on science and society; and with middle-class liberals celebrating the replacement of royal whims by parliamentary laws, it was perhaps natural that they should prefer Divine rule by *natural law* to capricious miracle. The socially-mobile God was promoted from feudal overlord to Divine legislator. Morse Peckham calls those advocating a lawful creation the "legalists of nature", a handy term and perhaps more appropriate than even he realised, since Owen actually drew considerable support from the Inns of Court next door to the College of Surgeons.

Owen needed a sensible alternative to transmutation embedded in a non-materialist framework, and he too turned to German transcendentalism, which he blended and muted with a liberal appeal to law. Far from the sterile hybrid that Huxley would have us believe, the union was astonishingly productive. First, it gave him the ideal Archetype, the "primal pattern" on which all vertebrates were based. This was a kind of creative blueprint, "what Plato would have called the 'Divine Idea'".[51] In practical terms, it was simply a picture of a generalised or schematic vertebrate; but this in itself provided him with a *standard* by which to gauge the degree of specialisation of fossil life, and in 1853 he saw it as an indispensable aid in determining the true pattern of emergence "of new living species".[52]

Although the Ideal Form was fleshed-out in increasingly special-ised guises according to the "predetermining Will", *the Archetypal pattern remained static*. Owen's secondary causes, metagenesis or whatever, were simply the means of translating the Word into flesh, the "Vertebrate idea" into the cavalcade of fossil life. There was never any question of what prompted the Idea in the first place,[53] and this set him on a profoundly different path from Darwin. Nonetheless, Owen's science was arguably more useful than Hux-ley's at this stage. If anything, the untethered bulldog was straying diametrically from his future Darwinian position, and grieving Darwin with his denunciations of progressive development. Hav-ing a measure of specialisation, Owen could now plot, say, the progressive departure of horses from some five-toed generalised antecedent to today's single-toed thoroughbred. In jettisoning the archetype Huxley lost this standard, and was loath to admit *any* evidence for fossil specialisation until well into the sixties, denying, e.g., that the Eocene *Palaeotherium* was in any way more generalised than the living horse (see Chapter 3 on Huxley's 'persistence'). So Owen's palaeontology was fruitful, and great caution is needed in assessing Huxley's post-Darwinian claim that his "mystifications" were for naught, when in fact it was Huxley who was forced to about-turn.

If Owenism was ever in vogue, it was from the late-1840s to the mid-1850s. Mantell might have come away from Owen's Royal Institution lecture "On the Nature of Limbs" in 1849 complaining that it was "too transcendental for any audience to understand",[54] but fellow "legalists" were more enthusiastic. Chambers had a morbid interest in hands, coming from a hexadactyl family and having been born with six fingers himself ("we are manifesting a tendency to return to the reptilian type", he joked to Owen), and the lecture on limbs struck him with "great wonder and interest". What he conspicuously neglected to praise, though, was Owen's advocacy of law, but then he was careful to avoid any subject which might expose him as author of the *Vestiges*.

The Reverend Baden Powell (1796–1860), Savilian Professor of Geometry at Oxford, showed no such reticence, being an out-spoken advocate of lawful creation and a cautious supporter of even Chambers' book. W. J. Hamilton at the Geological Society con-sidered Powell "one of the most philosophical writers of the day",[55] and he was spiritual mentor to a member of Huxley's circle, W. H.

Flower, which suggests an underlying continuity between post-Darwinian cosmology and this earlier theological tradition. Powell's radical stand for uniformity pushed him far to the left of Lyell and the mathematician Charles Babbage (of calculating machine fame). They would have agreed that our faith in "the immutability of the Divine attributes" rested on the "constant uniformity of the laws of the material world", but they refused to go as far as Powell and contemplate *transmutation*, imagining, like Owen, the operation of some less bestial mechanism.[56]

Powell championed Lyellian geology and urged men of the cloth to accept a "perpetual succession of creations". This raised daunting exegetical problems, but, not to be deterred, he bravely cut the Gordian knot by insisting that no tortured interpretation of Genesis would ever suffice. We had to let go the Days of Creation altogether and base Christianity on the moral laws of the New Testament. But this made it imperative that we seek God's Design in nature's uniformity. More than most, Powell realised that the expression "perpetual succession of creations" *was* theological, which made any attempt to found a natural religion on it absurdly tautological. Creation had to be explained *naturally*, with workaday causes, and therefore scientists had to be free to investigate any and every means of species origination. Only by establishing a continuous chain of causation back to Creation could we show Divine purpose in nature. Snap the chain – introduce a miracle – and the evidence vanishes, the design is lost and we are plunged into "the anarchy of chaos and the darkness of atheism". Fired by this need, Powell outflanked Lyell and Babbage and practically welcomed transmutation; at least, he gave it more than a fair hearing in an age that was prepared to dismiss it out of hand.

So there were deep-seated theological reasons why Powell stood so far to the left. But his resolve was strengthened as a result of being involved in a sectarian war with the Anglo-Catholics: the "semi-papists – the Puseyites, or *histrionic* Churchmen", as Mantell angrily called them.[57] The strength of doctrinal disputes should never be underestimated. The Anglo-Catholic movement was enjoying a considerable vogue in Powell's own university, and he hurled himself furiously against the Oxford Puseyites because they stood for those things he hated most, miracle and ritual. Also, Dr Pusey's denunciation of science had angered many naturalists and further made a devil's advocate of Powell himself. So it was partly in opposition to the "semi–papists" that he insisted geologists be given free rein; and he admitted composing his *Connexion of Natural and Divine Truth* (1838) as an antidote to Anglo–Catholic "mystical superstition".

This brings us back to shared interests. Powell contacted Owen in January 1850 after reading *Nature of Limbs* to express his high "satisfaction" at this "extended view of the argument for design from 'unity of plan'".[58] In reply Owen himself grumbled that he too had been attacked (in the *Manchester Spectator*), presumably by a young Puseyite priest who mistook his meaning. He said this unsettled his Lancashire relatives, and though he stamped on the "reptile" in a follow–up letter to the *Spectator*, he still welcomed Powell's antidote to its venom.[59] Powell was by now contemplating a new work and promised to take Owen's *Limbs* into account. These *Essays* on the "Philosophy of Creation" eventually appeared in 1855, and continued Powell's call for the *"universal and permanent uniformity of nature"*. He flirted ever more dangerously with transmutation, embedding Owen's archetype in an extended discussion of life's development, after noting that the French morphologist Geoffroy St Hilaire had seen "the possible migration or transition from one species to another" as a corollary of the "primary plan". Powell knew he was usurping Owen's archetype. He had actually tucked in a postscript to his letter asking Owen's view on development (i.e. transmutation), adding that he had heard no objections "so fatal to it as the objectors seem to imagine". It was a touchy subject and Owen replied by return. But rather than upbraid his friend, he enclosed a press clipping from the *Spectator*, his rejoinder to the Puseyite "reptile", emphatically denying any allegiance to the "stale" school of Transmutation. He did however add that if the evidence for a particular secondary cause producing a new species were as good as that for the archetype, his feeling of reverence for the "Cause of all Causes" would be appreciably increased.

Despite Owen's *caveat*, Powell pressed on, weaving the archetype ever more firmly into his developmental fabric; pointing out that Owen's ideas acquire their "highest interest" when considered in the light of the past progression of life, and deliberately quoting the famous passage closing the *Nature of Limbs* which had so outraged the Anglo–Catholics:

To what natural laws or secondary causes the orderly succession and progression of such organic phenomena may have been committed we are as yet ignorant. But if, without derogation of the Divine power, we may conceive the exist-ence of such ministers, and personify them by the term 'Nature,' we learn from the past history of our globe that she has advanced with slow and stately steps, guided by the archetypal light, amidst the wreck of worlds, from the first embodiment of the Vertebrate idea under its old Ichthyic vestment, until it became arrayed in the glorious garb of the Human form.

Quick to qualify this "noble passage", Powell insisted that,

not only "without derogation of the Divine Power," may we entertain the ideas so beautifully expressed; but . . . so far from anything *derogatory*, such a view constitutes the *very proof* and manifestation of that power . . .[60]

The archetype served Powell's purpose well, though its pious prostitution must have caused more than raised eyebrows. Nonetheless, Owen believed that secondary causes had ushered in new species, and Powell's sense of high design in an orderly cosmos was especially appealing, with its antecedent Intelligence and arche-typal Plan. Owen therefore welcomed the *Essays*, as much for their appreciation of his own contribution as their philosophic content, telling Babbage they were "to the point & timely".[61] Others tended to agree in a cautious sort of way (no other was possible with Powell broaching the inflammable subject of transmutation). Hamilton saw the subject as "the highest consideration of our science", and while baulking at the transmutation, on the grounds that fossils did not seem to follow in the necessary succession, he too recom-mended the *Essays* as deserving "a careful and attentive reading".

The moral purpose behind Owen's science is clear: to prove that Man was in the Divine Mind at the time of Creation. Owen knew of course that not all fossil lines pointed the human way, in fact only one of many did – still, there was a timeless purpose behind nature's veneer. Romanticism this was, though of a typically British variety: shadows of change masked an eternal truth, a preordained Plan. But Owen was never one to accept the panpsychic mysticism of the

German nature-philosophers, under the influence of F. W. J. Schelling, the Prince of Romantics. For Schelling nature was immanent in God and the Divine Intelligence reached out to express itself through a kind of cosmic poetry. Owen denied that the "great Cause of all" was "an all-pervading *anima mundi*",[62] the more pointedly, perhaps, because Schelling had actually pleaded guilty to a sort of pantheism, and Owen himself had been accused of it by the Puseyites. Rather, his God was a traditional British craftsman working to a blueprint. Nonetheless it all proved far too heady at home. Trying to drum up interest in a translation of *Nature of Limbs*, Owen admitted to Rudolph Wagner of Göttingen that the "subject is better adapted for the character of mind and thought of a German audience than for our matter-of-fact English".[63] But worse by far, his timing was inopportune. Transcendentalism was on the wane; younger positivists had begun to look on it as a bad mixture of pomposity and mysticism, and Huxley was not slow to see the capital in slating such "mystifications".

Rejecting the Archetype

Emasculation of the archetype was well under way by the fifties, which left Owen's palaeontological superstructure in 1859 begging for a Darwinian gloss. The demystified archetype doubled for Darwin's ancestor or "ancient progenitor",[64] and, more important, Owen's cases of fossil specialisation – the Tertiary horses and rhinoceroses – became prime examples of evolution. This is not to suggest that the transition was smooth. Ironically Huxley himself abhorred the notion of progressive specialisation in the 1850s and early 1860s, and would have denied Darwin the use of Owen's ready-made lineages – causing Michael Bartholomew recently to question his suitability for the post of "devil's disciple" (Chapter 3).

Significantly, Huxley's philosophy underwent little change during the watershed years 1855–1865. Of course Darwinism did make an impact. Indeed, the Secretary of the new Anthropological Society, Charles Carter Blake, was astounded at the speed of Huxley's about-face on the question of evolution. "There was a time, ten years ago," he complained in 1863, "when the same eloquence which is now employed in lowering 'Man's place in nature,' uttered its vehement strains in the scientific theatre of Albemarle Street in disparagement of the transmutation doctrine."[65] But this may not be the best way to gauge Huxley's position, for it ignores the continuity of his thought at a deeper level. We could say that he accepted Darwin for the reason he rejected *Vestiges*, which he denounced in 1854 as "a mass of pretentious nonsense" and "so much waste paper" – i.e. because Chambers had seen Divine edicts manifesting as nature's laws, which he pictured as spiritual movers behind the scenes. By confusing "law" with "cause", *Vestiges* had palpably failed to give any "*explanation* of creation", short of describing the fossil ascent in terms of an "orderly miracle". It wasn't the transmutation that Huxley objected to, and he probably only accepted it in Darwin's case because it was the corollary of naturalistic causes: population pressure, variation and selection. In other words, he never budged in his hatred of "pseudo-scientific realism", consigning Chambers' "laws" and Owen's archetypes to the same scrap heap as the Duke of Argyll's books thirty years later.

Not that he had found a *schematic* archetype unhelpful. In 1853, while he was still fired by Carlylean ideals, Huxley considered the archetype an indispensable aid to classification, going as far as to rate it "as important for zoology as the theory of definite proportions for chemistry".[66] But positivists violently objected to any idealist edifice being raised on the archetype. As Lewes said in 1852, "when it is no longer used as an artifice, but presented as an ascertained *plan* according to which the structures were composed – a *scheme* of creation subsequently realized, then indeed the positive philosopher demurs". Huxley tacked a similar proviso to the paper which Owen was expected to pooh-pooh. He used a diagrammatic representation of a mollusc, called it the "archetype", and explained that it made "no reference to any real or imaginary ideas", but simply summed up the most general propositions concerning molluscs and stood like a "diagram to a geometrical theorem". Darwin approved up to a point: "The discovery of the type or 'idea' (in *your* sense, for I detest the word as used by Owen, Agassiz, & Co) of each great class, I cannot doubt is one of the very highest ends of natural history . . ."[67] But for all practical purposes Huxley's archetype was as immutable as Owen's, being by definition incapable of self-change or progression, whereas Darwin's had long been

cast in an evolutionary light. In the back of his personal copy of
Nature of Limbs, Darwin had jotted: "I look at Owen's Archetypes
as more than ideal, as a real representation as far as the most
consummate skill & loftiest generalization can represent the parent
form of the Vertebrata"★ – in short, it was the best guess at what the
real ancestor had looked like. And obviously as an ancestor it had
the potential for self-improvement, hence Darwin's letter to Huxley in 1853 continued: "I sh^d. have thought that the archetype in
imagination was always in some degree embryonic, & therefore
capable & *generally undergoing* further development." The italics are
Darwin's (although missing from the printed version in *More Letters
of Charles Darwin*) and show where his emphasis lay. Another
important point must be made: in the fifties Huxley had no great
love for palaeontology, and therefore never placed his archetype in
an historical context – unlike Owen, who used it as a rationale for
the progressive specialisation of fossil life. As a result, only Owen's
could be replaced by Darwin's 'ancestor'.

Barring the odd jibe, Huxley resisted any headlong confrontation
with Owen's Platonism until his Croonian Lecture to the Royal
Society on 17 June 1858. Even then he neglected to mention Owen
by name – the least concession he could make with the man actually
sitting in the Chair. But that did not stop Huxley from castigating
archetypal terminology for defying the spirit of the age. Nor did he
leave anyone in two minds as to his target. As if to confirm it, he
dropped Hooker a note the following day, confessing: "My head
was more full of bones than brains. I wonder how Ricardus, 'Rex
anatomicorum', feels this morning."[69] The Croonian also encouraged others to speak out; for example, it sparked off Spencer's
blistering attack on Owen's idealism, the heat and smoke from
which tends to disguise certain deep-seated similarities in their
thought (Chapter 3).

It is too simplistic to see Owen's science castrated solely in the
name of secularism, whatever the naturalistic bent of the younger
writers and teachers. True, the Albemarle Street conspirators took a
moral stand on physical truth and made Uniformity their creed; and
they were to cold-shoulder anyone suspected of split loyalties in

---

★ Dov Ospovat, who transcribed this note, goes on to suggest that at this late date Darwin
still believed in a "created archetype", in the sense of having been produced by "designed
laws".[68] Indeed, he pictures Darwin as a theist until at least 1859, whereas others, most notably
Howard Gruber in *Darwin on Man*, would have him an 'agnostic' as early as 1838.

their push to create a self-governing scientific civil service. But a number of Christians shared their professional ideal and were equally opposed to Owenism. For them religious experience was a more private affair and unlikely to intrude on their chosen career. And by and large they found the transition to Darwinism relatively easy, inasmuch as it did little violence to their deeper beliefs. Only in Parker's case, where there was liable to be a clash, was there any kind of enforced separation or schizoid thinking.

The Christian Commitment

William Kitchen Parker (1823–1890) was something of a theistic counterpoint to Spencer. Spencer's urban upbringing among the Wesleyan manufacturers of Derby led to his political radicalisation and loss of faith. Parker by contrast was born into a Methodist farming community fifty miles away in Dogsthorpe, near Peter-borough, and his life-long almost rustic piety was reminiscent of Faraday's. An "abiding sense of the Divine presence" coloured his every moment, his son records. He apparently suffered no crisis of faith, and looking back late in life had only scorn for the infidel:

> the supposed old wives' fables and cunningly contrived deceptions of the four gospels, which according to —— it is immoral to believe, these supposed inventions of those old Jews have been the strength of my life – they have lifted up my mind. As for the knowledge of modern science, it is only ignorance with its eyes open.[70]

A metaphysical abyss separated Parker and Huxley (whose name could easily be inserted in the gap above). Yet their friendship was firm and enduring, with Huxley acting as a sort of literary midwife, critic and referee.

Parker's fate, as a comparative anatomist seeking work in London, was closely bound with Huxley's and Flower's. He had entered Charing Cross Hospital in 1844, while Huxley was still a student, and attended Owen's lectures at Lincoln's Inn, like so many at the time becoming enthralled with the archetype. But it was not to last. Parker carried out a number of meticulous studies of skull development, so embarrassingly detailed that he was awarded a Royal Medal in 1866, and it was this work, close to the heart of Owenism, that finally shook his "faith in the great system of transcendental anatomy". His painstaking work helped to crush Owen's theory of the vertebral skull (the belief, common at the time, that the skull was composed of fused and modified vertebrae). Since Huxley in his Croonian had singled out the same target, he invested considerable time and effort in improving Parker's papers. Parker by now had completely swung into Huxley's camp, and in one paper he deplored the anatomical suffering caused to fish

> from their being dragged into harmony with that mischievous piece of fancy-work, "the vertebrate archetype." It is high time for us to have ceased from transcendentalism: of what value is it? Our proper work is not that of straining our too feeble faculties at system-building, but humble and patient attention to what Nature herself teaches, comparing actual things with actual.[71]

Parker's exuberant belief in Old Testament miracles may well explain why he, of all people, lined up with the sceptics. Holding to the miraculous forced him to separate science and theology into "watertight compartments". Asked to join the Victoria Institute, for instance, he declined on the grounds that reconciliation was impossible: "Let your society try and explain the swimming of Elisha's axe-head by the laws of hydrostatics; I believe the axe-head swam, but I don't believe it can be explained." In other words, science was impotent in the face of the miraculous, but equally theology was no guide in mundane methodological matters. Hence the ease with which he could accommodate a positivist biology, borrowed from Huxley, and which justified him scrapping the archetype.

More commonly, Anglicans who rationalised their faith and regarded Mosaic miracles as allegory still sought a scientific basis for natural religion – to the extent that *theological* imperatives could account for the destruction of Owen's archetype. Such certainly seems the case for William Henry Flower (1831–1899), friend to both Huxley and Parker, but again a man deeply dismayed "by the combative character of some of the then leaders of science and best equipped exponents of evolution."[72] Flower was the son of a wealthy brewery owner and a likeable Broad Churchman in the Baden Powell mould. Powell was actually his wife's brother-in-law and the family connection had been close. Hence Flower's early exposure to the theological need for a wholly natural creation. Perhaps this explains why he was later singled out by the Church Congress to explain the intricacies of evolution, a lecture unfairly sent up by Huxley as an attempt to lure the "ecclesiastical swallow" with a well-oiled bolus of evolution. Flower's actual address was noticeably more sympathetic. Evolution had not altered one whit the "evidence of the Divine government of the world":

> The wonder and mystery of Creation remain as wonderful and mysterious as ever . . . It is still as impossible as ever to conceive that such a world, governed by laws . . . could have originated without the intervention of some power external to itself. If the succession of small miracles, formerly supposed to regulate the operations of nature, no longer satisfies us, have we not substituted for them one of immeasurable greatness and grandeur?[73]

One-time co-worker Richard Lydekker took this to mean that Flower was uncomfortable with Darwin's theory of the survival of the fittest. But the opposite is probably true. Not only did Flower testify to his own Darwinian orthodoxy, but the real point is that in his Powellian view natural selection made creation rational; without this mechanism it was inexplicable, or rather, reduced to a string of minor miracles. What makes this interpretation compelling is Flower's approach to Owenism. He had probably been reared on Owen's *Archetype*. (The evidence is circumstantial: he had studied at University College, taking Sharpey's Gold Medal in physiology and Grant's Silver in comparative anatomy in 1852, and according to Spencer Owen's book was a course text in Gower Street at the time.) But Flower came to reject it. The "*type* or *common plan*"

explained nothing, accounted for nothing. It gave not even a shadow of a reason for the resemblance amid diversity found everywhere. It only required that the Creator had imposed certain apparently quite arbitrary restrictions on His power, but, beyond this almost paradoxical assertion, it gave no clue to elucidate anything like a theory of creation.[74]

But a theory of creation is precisely what Darwin had given, to Powell's express delight, and Flower, like Powell, probably accepted Darwinism for its more sublime view of Divine government.

Broad Church biologists could still insist with Huxley that the archetype defied the spirit of the age. George Rolleston (1829–1881) is a case in point. A vicar's son from Yorkshire, he had read Classics at Pembroke College in the forties, before gaining a medical Fellowship and moving to St Bartholomew's Hospital. During his student days he spurned the shrouded symbolism of the Oxford Movement and, influenced by Huxley in London, rejected the mystical garb of Archetypalism. Huxley evidently thought enough of Rolleston to support him for the Linacre Chair of Anatomy and Physiology at Oxford in 1860. A week later it was heard that Owen had armed his rival, Waller, with a "very strong" testimonial, and, Rolleston told Huxley, "It has been hinted to me that some of this strength is due to a feeling of antagonism on his part to you."[75] Rolleston was successful and pledged Huxley "never to give you cause to regret the share you have had in my promotion." However much "Oxford slough" Rolleston still had to shed, Huxley evidently found him "plastically minded" enough to consider enticing into co-editorship on the *Natural History Review*, even while boasting to Hooker that the new series would be "mildly episcophagous". Although Rolleston never joined the team, he did carry out missionary work in Oxford, trying to convince friends of the ease "in reconciling Darwin with the Established Creed", while publishing a number of pro-Huxley papers in the *Review* and pointing up "the utter incongruity which exists between Platonic mysticism and modern science".

Revealing the disfigured body hidden by the archetype's "antique dress", Parker, Flower and Rolleston were not covertly hitting the Church, or attempting to usurp its earthly power. Perhaps the reverse – each could have had theological reasons for rejecting Platonism. Of course, it was this hatred of *archetypalism* which explains the fact – incongruous within the 'Science *vs.* Religion' paradigm – that Huxley, the self-proclaimed smiter of Amalekites, had a large hand in promoting all three men to their professional posts. Fellowships of the Royal Society came in quick succession. Rolleston gained his in 1862 at the height of the ape-brain debate (next chapter), both he and Flower having already come out on Huxley's side against Owen. Flower's followed in 1864, after Huxley had sought Darwin's blessing ("I go bail for his being a thoroughly good man in all senses of the word"),[76] while Huxley put up Parker's name a year later, helping him on his way to a Royal Medal for establishing "the true theory of the vertebrate skull". The timing and political importance of these moves should not be underestimated. Huxley saw acceptance in Society circles as tantamount to official recognition of one's science, and this made it crucially important, not only to get his own men into key positions, but – we will see – actually to veto Owen's appointment to the Council.

Rolleston talked of the failing empire of "Archetypal Ideas", and so it must have seemed with the abrupt transfer of power at an institutional level; indeed, nowhere was it more astonishing than in Owen's own College of Surgeons. After two decades as Hunterian Professor, Owen had made Hunter's museum one of the most prestigious in the country. Here if anywhere Owen's thought should have endured, if only because he could have influenced, as no one else, the choice of successor. But apparently not. He clashed with the Council and exited in 1856, only to see a succession of Huxley sympathisers move in with indecent haste. Flower first; backed by Huxley, he was unanimously elected Conservator in 1861, though the youngest of fifty candidates (among whom was Parker). And only months later (in 1862) Huxley himself took the Hunterian Chair, a jubilant Flower informing him that April that the Court of Examiners had

> unanimously decided that "there's not a man in all Athens that can discharge Pyramus" (*i.e.* the Hunterian Professorship) but you . . . I am exceedingly rejoiced myself at the prospect of the new "Hunterian Professor," though I don't know what *our* illustrious predecessor will say.[77]

So within six years of Owen's departure, the Museum had switched from the bastion of English transcendentalism to a bulwark of anti-Platonism. And so it was to stay.

2

Creative Continuity: Fossils & Theology

The Trouble with Retrospectives

In his *Edinburgh* review of the *Origin*, Owen denied having any "sympathy whatever with Biblical objectors to creation by law, or with the sacerdotal revilers of those who would explain such law".[1] But the review was otherwise so mealy-mouthed that his denial has frequently been taken as another jesuitical aside. The irony is that for once Owen was being quite candid, although the fact was lost as Huxley in a rival *Westminster* review painted him one of "the most decided advocates of the received doctrines respecting species". This indictment proved so effective that only a generation ago Loren Eiseley could lump Owen and the glacial cataclysmist Louis Agassiz together as "old style catastrophists and progressionists" and supporters of "separate creations".[2]

This is the danger of accepting propaganda at face value: history has come to perpetuate and sanctify the partisan view. The original need for that view is easy enough to understand. The early evolutionists doubled as historians and did what any nationalistic group does to forge its self-identity – they wrote the kind of heroic history encouraged by Victorian positivism, portraying the forces of light triumphing over religious obscurantism. Corpses were disinterred and the once-vanquished ceremoniously reslain, something very much in evidence in Darwin's and Huxley's *Lives and Letters* (where enemies like Owen are consistently vilified). These propagandist histories are perfectly adequate for their purpose, so long as we remember that this *is* history written from the 'victor's' viewpoint.

Huxley was so engrossed in his military-style campaigns and so swept along by contemporary social currents that his reminiscences are understandably unidimensional. Nowadays, though, historians are no longer content with reslaying the slain. They want to see beyond partisan bounds; they need to *sympathise* with the positions of other parties in order to frame a larger canvas. So in this chapter I will try to distil out the propaganda, and in the process show why there was a tactical advantage to Huxley in painting Owen a reactionary. This should clear the way for a more reasonable picture of Owen's palaeontological world view.

The first and most obvious point to note is that in the 1850s – the time he was looking back on from his post-Darwinian vantage point – Huxley himself was no palaeontologist. Offered Edward Forbes' posts of Palaeontologist and Lecturer in Natural History at the School of Mines in 1854, he actually "refused the former point blank" because he "did not care for fossils".[3] Indeed, before 1859, Huxley devoted little time to palaeontology, just enough to describe a new crustacean from the "Coal-Measures", and to point out that *Pteraspis* was not a fossil squid but probably the oldest known fish. Only when he saw the polemical importance of fossils did he exploit this resource in any systematic way, first (in the later 1850s and early 1860s) using archaic amphibians and persistent crocodiles to destroy the case for geological progression, and later – changing face somewhat – employing dinosaurs and the same crocodiles to establish genealogies in aid of Darwinism.

Not that the young professor was bashful of lecturing old hands on palaeontological procedure. In 1856, before publishing a single descriptive paper of his own, he gave more experienced men qualms by subjecting the century's foremost comparative anatomist, Georges Cuvier, to strenuous criticism in a Royal Institution lecture. Hugh Falconer, a respected palaeontologist recently returned from India, was absolutely furious, and Darwin knew that Owen would be outraged. Understandably so, for by this time Owen had authored over 100 books and papers on vertebrate palaeontology: monographing British fossil reptiles and mammals, christening the dinosaurs, and making a start on colonial imports – reconstructing the New Zealand moa and pioneering the study of South African mammal-like reptiles and Australian Pleistocene marsupials. With this track record he was obviously unhappy at having a tyro telling him how it was done. But then Huxley just revelled at the prospect, telling Dyster on 3 November, "There is going to be a set-to at the Geological on Wednesday. The great O. versus the Jermyn St Pet on the methods of Palaeontology. You shall know the results."[4] The iconoclastic aspect of the paper worried even Darwin, usually the last with a reprimand, and caused him to reconsider proposing Huxley for the Athenaeum Club, holding off for a couple more years in the hope that he would mellow.

Huxley's lack of acquaintance with practical palaeontology in the 1850s, coupled with his later commitment to Darwinism and positivist historiography, go some way to explaining his often distorted reminiscences. To pick the most blatant case – in "The Coming of Age of the Origin of Species", which he placed in *Nature* in 1880 to celebrate the *Origin*'s Twenty-First (itself a political act, with Darwin pictured receiving the key to Science's door), Huxley wanted to create the image of an abrupt about-turn in 1859. He achieved this by reinterpreting geology before the *Origin* as catastrophic. "Great and sudden physical revolutions, wholesale creations and extinctions" were the norm, as they had been since Cuvier's day. But as numerous historians have shown, catastrophism had waned well before the fifties, and a careful reading of the papers and Presidential addresses of the Geological Society for the 1850s demonstrates how false Huxley's memory played him. True, Agassiz at Harvard was still sending ice-sheets sweeping across the globe "like a sharp sword" to sever past and present life. But Sedgwick in the 1830s had despaired of such "moonshine", and Darwin thirty years later thought Agassiz "glacier-mad"; even Huxley in 1854 dismissed him as creeping towards the lunatic fringe.[5] Élie de Beaumont in France also continued to insist on paroxysmal crustal collapses into the 1850s. Yet every Londoner who "had sucked in [Lyell's] *Principles* with his mother's milk" was safely immunised. When de Beaumont produced a technical three-volume work in 1853 which proposed a mechanism for these crustal wrenchings, Forbes rearranged his Presidential address that year to dispose of it.[6] Owen was no field geologist himself and leant heavily on Lyell for his gradualist views, and in the Introduction to *British Fossil Mammals* (1846) admitted that fossil jumps are best explained by the weathering out of intervening strata. Lyell wrote in a paean of praise:

. . . I have read nothing of late which has given me more pleasure. It comes out most opportunely together with E. Forbes' beautiful essay on the Flora & Fauna & the identity of the crag shells with living ones to combat Alcide D'Orbigny & Agassiz with their division "as sharp as a sword" between formations & groups of species & the non-passage of fossils from one set of strata to another. You have brought out the absence of a catastrophe well & clearly & I think you never wrote anything more tersely as well as in good style.[7]

No London geologist in the 1850s seriously believed that crustal convolutions signalled epochal changes. So Huxley was not writing history, but propaganda which turned the past to advantage. His stratagem was to relocate the extremists at the centre of the stage and present a palaeontological contrast: the catastrophists with their "great breaks", and the Darwinians with their "intermediate grada-tions". This allowed him to ask the obvious question – how was it that Darwin carried the world with him *after* 1859? Huxley answers with a survey of subsequent palaeontological progress, gingerly stepping round Owen, and outlining his own work on *Archaeop-teryx* and the dinosaur-bird connection, while concluding that the "whole tendency of biological investigation since 1859" has been to close the fossil gaps. With such success, he adds, that "if the doctrine of Evolution had not existed Palaeontologists must have invented it."[8]

But as we will see, a fine study of Owen's work on fossil reptiles suggests that all is not so straightforward. In fact he had already made an effort to close a major gap – between fish and reptiles – not only before Huxley's attempts at genealogy, but even before the *Origin* was published. As a consequence, I would suggest that to some extent Owen's palaeontological views, far from clashing with Darwin's, were actually supportive.

Expectations

The Lubbocks, Father & Son (God save the mark) thought it was Owen's!!!

> Darwin referring to the
> anonymous *Times* review of the *Origin*.[9]

This brings us to how Owen was expected to react to the *Origin*. For all of Huxley's exaggerations, it was known that Owen was no bibliolater; in some circles it was expected that his belief in "continuous creation" would actually predispose him to evolution. Not everybody, of course, went so far as the Lubbocks as to see him writing a rave review in *The Times*. As it happened, Darwin had a more realistic insight into Owen's psychology. The appreciative *Times* review appeared on Boxing Day 1859, and Darwin correctly guessed that it was Huxley's doing, telling him:

> Upon my life I am so sorry for Owen; he will be so d—d savage; for credit given to any other man, I strongly suspect, is in his eyes so much credit robbed from him. Science is so narrow a field, it is clear there ought to be only one cock of the walk![10]

A prophetic forecast, for Darwin was to be hauled over the coals in Owen's *Edinburgh* review in 1860. All this renders Owen's pleasantness in December 1859 a little suspect; meeting Darwin that month, he apparently told him that the newly-published *Origin* contained the best explanation "ever published of the manner of formation of species". Now, we know that Owen had grave reservations about transmutation, preferring an alternative non-transmutational mechanism. So how are we to explain his unaccustomed generosity? Well, if Ospovat is right, and Darwin was still a theist at this stage, perhaps he told Owen what he was to tell the Harvard botanist Asa Gray, viz. that he was "inclined to look at everything as resulting from designed laws . . ." From this Owen might have gathered that they shared a certain common ground; it might also have looked as if their systems were formally similar in some respects. Owen could *afford* to be polite, leaving Darwin with the distinct impression that "he goes at the bottom of his hidden soul as far as I do!"[11] Perhaps Darwin even read his own evolutionary inclinations into Owen's lawful explanation of fossil life. Whatever the truth, the two men seemed reasonably pleased with one another. So why Owen's change of heart? Why the uncharitable *Edinburgh* review? It was probably nothing so crude as treachery or deliberate deception, as older histories consistently imply. Owen possibly only realised the real danger on seeing Huxley take up the

Darwinian cudgels and ruthlessly exploit evolution for materialist ends, bestialising man in the process. Suddenly, the potential threat of Darwinism in hostile hands, in *Huxley's* hands, would have become frighteningly apparent.

There is no doubt that Owen accepted a causal explanation of species before 1859. We also have St George Mivart's word for it. (Mivart, part taught by Owen, part by Huxley, was a liberal Catholic and skilful anatomist, whose dissections of the lemur were used by Darwin in the *Descent of Man*.) In later years Mivart gravitated back to Owen, considering his archetypes the more important for being "refreshed and invigorated by the bath of Darwinian Evolutionism through which they have been made to pass". Having nailed his Platonic colours to the mast and become caught up in a pamphlet war with the American Darwinian Chauncy Wright in 1872, Mivart vowed to do Owen "a simple act of justice", namely to bear "witness to your distinct enunciation of the evolution of species by the operation of secondary laws before ever the 'Origin of Species' saw the light."[12] Again we must beware, because Mivart was using hindsight: "evolution" is not a word Owen would have employed in the 1850s. Nonetheless, this episode suggests that Owen had discussed the *lawful* origination of life, convincing Mivart years later that he had cornered evolution first.

By all accounts, the bibliolater image has to go, and I think we might reconstruct a more subtle triangular relationship between Owen's palaeontology, theology, and science of species. In short, I want to suggest that he had reached a position which the *Athenaeum* called "continuous creation".[13] It was formally similar in palaeontological respects to Darwin's "descent", and rested on fossil evidence which would have served Darwin equally well.

The New Design

The axiom of the continuous operation of Creative power, or of the ordained becoming of living things, is here illustrated by the class of fishes . . . But the creation of every class of such animals, whether Reptiles, Birds, or Beasts, has been

successive and continuous, from the earliest times at which
we have evidence of their existence.

> Owen, speaking as President of
> the British Association, 22
> September 1858.[14]

In 1858 Owen was at his professional peak, even if young bloods
like Huxley and Carpenter were already roaring at his "absurd-
ities". That year saw him elected President of the British Association
at Leeds, while already Fullerian Professor of Physiology at the
Royal Institution, lecturer at the School of Mines, and Superintendent of the Natural History collections at Bloomsbury. This gave
him a platform of unprecedented power and he exploited it in his
public crusade for a National Museum of Natural History.[15] Towards this end he had begun making political allies. In May 1858 he
was invited to Gladstone's Thursday morning breakfasts, and it
was Gladstone,[16] as Chancellor under Palmerston, who introduced
a bill in 1862 to push through Owen's plans for the museum
(unsuccessfully at the first attempt). This period was perhaps the
closest Owen came to being a spokesman for science, his impeach-
ment by Huxley in 1860–3 for "wilful & deliberate falsehood"[17]
effectively scuppering any intention he might have had of being an
elder statesman of science. So it is essential to look at what Owen
was saying in 1858, on the eve of the *Origin*, when he was still in a
position of influence.

In his 1858 address to the British Association and concluding
Fullerian lecture "On the extinction of species" (1859), Owen made
a philosophic distinction which affords crucial insight into his
appreciation of the status of scientific laws. By "patient and inductive research", he claimed, the natural philosopher has established
certain unassailable axioms, never to be confounded with the
"speculative philosophies" invented to explain them. In other
words, natural laws were divinely instituted and inviolable, whereas explanatory hypotheses were a product of human ingenuity. He
had in mind the "axiom of the continuous operation of Creative
power" as an inductive truth. This could be read off from the fossil
record, which revealed a progressive incarnation of the Ideal
Archetype. Quite distinct, however, were the various hypothetical
mechanisms: transmutation, metagenesis, or whatever. This distinction was trotted out time and again – at the British Association,

Royal Institution, in his *Edinburgh* review of the *Origin*, and in *Palaeontology* (1860), where he states:

> As to the successive appearance of new species in the course of geological time, it is first requisite to avoid the common mistake of confounding the propositions, of species being the result of a continuously operating secondary cause, and the mode of operation of such creative cause. Biologists may entertain the first without accepting any current hypothesis as to the second.[18]

Romantic liberals regarded this separation as not only reasonable but crucial, and the Duke of Argyll reassured Owen:

> You are quite justified in drawing a broad distinction between "Creation by Law" – (*some* Law) – and the special theory in respect to the precise nature of that Law, which Darwin has put forward. At the same time "Creation by Law" is necessarily connected in our minds with "*Creation by Birth*", Generation being the only "Law" which we know of as capable of "Creating". The Law under which a *new Birth* takes place is very vaguely shadowed forth in the Theory of Darwin. But he has probably made a guess which includes a few *bits* of the truth – nothing more.

Yet to the Huxley faction with its mechanistic and secular needs this appeared a useless and perhaps pernicious distinction, and Huxley himself viewed Owen's *ex cathedra* pronouncements with disdain. Unfortunately, Owen was his own worst enemy, and his use of awkward expressions practically invited Huxley's sarcasm. Owen, for example, referred to "*the continuous operation of the ordained becoming of living things*", which made little sense without knowing his commitment to Powellian design and advanced views on fossil specialisation. To the swelling ranks of positivists, rationalists, and later, Darwinians, his words seemed quaintly archaic, if not laughably evasive. Looking back to *Nature of Limbs* from his Darwinian vantage point, Flower thought Owen's "trumpet gave an uncertain sound", but it was Huxley who extracted the greatest capital in his *Man's Place in Nature*:

> though I have heard of the announcement of a formula touching "the ordained continuous becoming of organic

forms," it is obvious that it is the first duty of a hypothesis to be intelligible, and [this] . . . may be read backwards, or forwards, or sideways, with exactly the same amount of signification . . ."[19]

Of course only Owen is to blame for the semantic pandemonium, and he must have intended a certain ambiguity; indeed, it was a useful cloak, protecting him from the attacks of conservatives, especially the Puseyites. Owen edged forward hesitantly, desperate lest he offend, influenced by Powell's "timely" *Essays* but acutely conscious of the reactionary ferment whipped up by *Vestiges*. He evidently chose his words carefully so as not to prejudice his own case, even if it left him making little sense. His tactics can be likened to Darwin's "truckling", when in the "Transmutation Notebooks" he ruminated on how best to disguise his materialism to gain a fairer hearing, and in the *Origin* talked of the Creator "breathing" life into primordial forms.[20]

Many of the moral objections to "ordained continuous becoming" Powell himself had cleared up, and Owen plotted a particularly Powellian course. He assuaged fears by placing organic science on a philosophical and moral basis akin to that of natural philosophy. Like Powell in his *Order of Nature* (1859), Owen looked to the *"grand principle of law"* as the basis of Omnipotent design.[21] Just as Newton established God as the "Lawgiver" in the physical realm, so the axiom of "ordained becoming" in a "creation [that] has been successive and continuous" established Him as lawgiver in the organic realm. The Almighty need exert His presence by periodic interruption no more in the creation of species than in the direction of the wind. Both were subject to laws by which He preordained the course of events. To Owen the idea of a progression of well-adapted species in ever-changing environments was nonsensical without an *"anticipating* Intelligence", and on this he was in one mind with Babbage, Chambers and Powell. Owen's hope was possibly to carry more orthodox scientists with him, and in so doing sweep aside metaphysical objections to lawful creation; hence he proudly concluded that in establishing a lawful biology "our science becomes connected with the loftiest of moral speculations". In the ensuing reconciliation of Christianity and evolution such a reformulation was to prove decisive.

Archegosaurus – Closing the Fossil Gap

Since the discovery of the Pterodactyle, probably no event in the domain of palaeontology has been more important than the discovery of the Archegosaurus.

> The Frankfurt palaeontologist
> Herman von Meyer, 1848.[22]

Archegosaurus conducts the march of development from the fish proper to the labyrinthodont [amphibian] type . . .

> Owen reporting to the British
> Association in September 1859
> on the Carboniferous fossil
> which spanned two classes.[23]

However, Owen was no moralist but a palaeontologist, and we need to know how far he reformulated his geological doctrines to bring them into line with Powellian theology. The fossil record itself could be made to fit almost any prevailing theory. Hence in 1841, when it was the fashion to use fossil gaps and retrogressive sequences to thwart the Lamarckians, Owen insisted that reptiles had *not* emerged in fish-like simplicity, and that the oldest still resembled well-developed monitor lizards of today's Tropics. Even the Triassic labyrinthodonts were too crocodilian-looking to support a direct connection with the fishes. So he started off adamant that any sort of continuity or "hypothetical derivation of reptiles from metamorphosed fishes" was nonsense, and "directly negatived" by the evidence.[24] But he had already begun to reconsider by the late 1840s, as the Grantian threat receded and he gained confidence in his Archetypal theology with its Divine Plan realised in a streaming of fossil life. In the *Nature of Limbs* he accepted life's "slow and stately steps", and as his commitment to continuity grew in the 1850s he became receptive to 'transitional' fossils – those like *Archaeopteryx*, which he described in 1862 and which were so important from a Darwinian point of view. But before Darwin had published or *Archaeopteryx* was heard of, Owen had employed another fossil in exactly the same way – as a bridge between classes. This was the Carboniferous *Archegosaurus*, an archaic amphibian, about a metre long and with jaws designed for fish catching. The name means "primeval reptile" and was coined in 1847 by the Bonn professor Georg August Goldfuss in recognition of the animal's great age and lowly position. Owen was to consider it "remarkable", and the gifted German amateur von Meyer called it the most important find for half a century.

Unlike its evolutionary counterpart *Archaeopteryx*, the fame of *Archegosaurus* was short-lived and it ceased to steal the limelight after 1860. If anything, this makes it more intriguing, and the role Owen accorded it – as a kind of missing link, an "annectant" form he called it – tells us a good deal about his metaphysical imperatives. Few actually followed Owen – I can think of only Albert Gaudry in Paris, whose main aim was not to promote Darwinism but establish a close-knit fossil sequence. So by all accounts *Archegosaurus* might not have been the best fossil for the job. Huxley, when he eventually came round to the idea of evolutionary progression in about 1867, ignored it altogether; and for good reason, since in his famous Royal Institution lecture on intermediate fossils, he implied that only an *evolutionary* ideology could provide the proper incentive to fill the "great gaps" in the fossil record.[25] He could scarcely jeopardise his position by admitting that Owen had set the precedent, motivated by a rival ideology. (No doubt Huxley considered Owen mistaken in his use of *Archegosaurus*, but the point surely is that he had searched for intermediates in the first place.) This makes it imperative that we look at Owen's position afresh.

Opinion in Germany was originally divided on *Archegosaurus*. Incomplete skulls had turned up in Carboniferous beds around

Skull of the Carboniferous amphibian *Archegosaurus*. (From Goldfuss [1847].)

Saarbruck (near the border with France), which had previously yielded nothing higher than fish. In 1844 von Meyer was unable to diagnose these skulls and consulted the expert Louis Agassiz. In 1847 better specimens displaying limbs and toes were mono-graphed by Goldfuss, who christened them *Archegosaurus*, and von Meyer now considered that they were in fact the oldest-known labyrinthodonts. Labyrinthodonts were large flat-skulled amphibians, often known from little more than their disarticulated heads. Hence Owen restored them lifesize in the grounds of the Crystal Palace at Sydenham in 1854 as gigantic frogs and toads, whereas Huxley opted for a salamander shape. Size was a problem. Labyrinthodonts were generally huge; the *Mastodonsaurus* skull alone dwarfed the entire archegosaur. Nonetheless, von Meyer saw distinct similarities in the broad head, "weak" limbs and tiny feet "serving only to swim or creep".[26] Something too that Owen noticed, the *Archegosaurus* specimens showed traces of gills, or rather their branchial supports – which meant either that these were larvae or else that adults retained gills. Both Goldfuss and von Meyer concluded that these archaic water-dwellers had indeed lived in a "permanent Larva-condition".

This was the situation when Owen examined the "problematical" fossil. He had actually been given an original split nodule containing an archegosaur by that indefatigable collector of fossil fishes, the Earl of Enniskillen. Enniskillen was part of a European exchange network set up by enthusiasts, as a result of which he built up a sizeable collection at his country seat, Florence Court in Ulster (where most of his type specimens were described by Agassiz). He kept up a voluminous correspondence with Owen, dispatching fossils and intelligence during his travels, and it was while examining museum collections in Bonn that he had come across Goldfuss's *Archegosaurus*. "They have several heads," he reported back to Owen. "I find I have one at F. Court which I had got for a fish. I have written to my Curator . . . to send it to you."[27] This was Owen's only original, but according to his *Descriptive Catalogue* of Hunterian fossils (1854), he was also working with eight coloured plaster casts supplied by Goldfuss.*

---

* None has survived to the present day. The Hunterian Museum was bombed in 1941 and the collections sustained serious damage. The present Curator, Miss Elizabeth Allen, suspects that the *Archegosaurus* specimens were destroyed at the time.

Owen agreed that the skull roof resembled a labyrinthodont's. And although Goldfuss's casts were not so good as to show the "smooth and finely wrinkled" skin or "traces of minute scales" of the originals, the branchial supports were plainly visible in two specimens.[28] But Enniskillen's original, consisting of the skull and a portion of the trunk of a smaller "species", *A. minor*, enabled Owen to take issue with von Meyer: the skull, persistent gills and short ribs all pointed to a position between labyrinthodonts and sinuous eel-like amphibia such as the Congo Eel. So by 1854 Owen had yet to make *Archegosaurus* a link between fish and reptiles, even if he had lowered its position immeasurably. What is fascinating is how much further he was prepared to go. And having reached a definitive position five years later, how forcefully he was to insist that *Archegosaurus* had demolished the "boundary between the classes *Pisces* and *Reptilia*", almost as though fossil continuity were now a philosophical imperative.

After examining 271 specimens, von Meyer published an exquisitely-detailed memoir on *Archegosaurus* in 1857, and was rewarded with the Geological Society's Wollaston Medal in 1858.[29] Most astonishing was his discovery that the archegosaur's vertebrae were incomplete and surrounded a flexible notochord (an embryonic retention for sinuous swimming). In this respect, Owen saw "the old carboniferous reptile" resemble lung-fish like *Lepidosiren*, and he was now willing to argue that the archegosaurs and lung-fish obliterate "the line of demarcation between the Fishes and the Reptiles." When he lectured on "Fossil Birds and Reptiles" to an appreciative audience at the School of Mines in March 1858, he apparently concentrated on six major features linking *Archegosaurus* to lung-fish and suggested that fishes and reptiles formed "one great natural group" of unparalleled spread, and that the "blending" of the two groups made it practically impossible to draw a dividing line. Or rather, he saw the line as somewhat "arbitrary", a point he made strongly in his "Report" to the British Association in 1859.[30]

But he did more than advocate the "linking and blending" of two classes into one great serial development. Indeed, thinking of it as a *serial* development probably does him a disservice. For each of these "annectant" forms was the starting point for a divergent lineage. In short, Owen implicitly accepted the idea of 'branching', announcing that:

Archegosaurus conducts the march of development from the fish proper to the labyrinthodont type; *Lepidosiren* conducts it to the . . . modern batrachian [amphibian] type. Both forms expose the artificiality of the ordinary class-distinction between *Pisces* and *Reptilia*, and illustrate the naturality of the wider class of cold-blooded vertebrates . . .

Only convention made one a reptile and the other a fish, an artificial division which belies their real closeness. And it is to the air-breathing poorly-ossified lung-fish that we must go in search of amphibian roots, whereas reptiles originated in archegosaur stock. We can represent this diagrammatically, bearing in mind that Owen did no such thing (indeed this sort of diagram is decidedly post-Darwinian) – nor do I want to press Owen's claim too hard, since the idea of 'branching' has often to be coaxed out of his thought. Nonetheless a diagram will give some idea of the readiness of his palaeontology to take an evolutionary gloss.

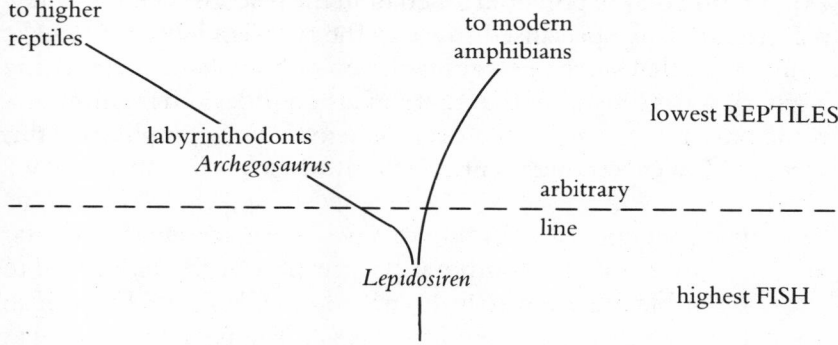

Peter Bowler remarks in *Fossils and Progress* (1976) that Owen's image of life was "now a highly sophisticated one, far closer to the modern approach than the old linear progressionism", and "in many superficial ways" his conclusions "were in agreement with the new evolutionism" ushered in by Darwin.[31] What the American H. F. Osborn later called 'adaptive radiation', i.e. the image of life branching and specialising, seems to have been implicit in Owen's thought before 1859. No longer was nature "straight as an arrow", as the Scots geologist Hugh Miller wrote in *Testimony of the Rocks* (1857). Paradoxically, it was possibly because Owen now saw fossil lineages develop by lateral specialisation, rather than stretch upwards towards man, that he could shift to genetic continuity without implying transmutation. After all, the London Lamarckian Robert Grant (see Chapter 4) still relied on a linear 'upward-striving' model of fossil progression (and as Swiney Lecturer on "Palaeozoology" between 1853–7, he was now openly arguing for ascent by natural "generation"[32]). But Owen's radically innovative picture of a *spreading* fossil record effectively immunised palaeontology against the potential Grantian threat, and left him free to adopt fossil continuity while insisting on some non-Lamarckian mechanism. Although the irony, of course, is that by removing to secure ground, Owen unwittingly played John the Baptist to Darwin's Christ.

Owen's theoretical reorientation was essentially complete by the late fifties. In his 86-page round-up of "Palaeontology" for the *Encyclopaedia Britannica* in 1859, he quietly stated that "the genetic history of fishes imparts the idea rather of mutation than of development".[33] By "mutation" he meant an explosion of adaptive forms all based on the same ground plan: variations on a theme, an exploitation on the potential inherent in the piscean type. Very few fish had actually specialised towards the reptilian level. In fact, he pointed out that since the main radiation of bony fishes occurred in Tertiary times long after the emergence of reptiles and mammals, it could have no relevance whatever to them. Bowler emphasised this aspect of Owen's thought and Martin Rudwick wrote in a review:

> It is particularly satisfying to have a clear account of how far the model of the temporal progression of life had come to resemble the irregularly-branched "tree" of post-Darwinian theorizing, even *before* Darwin's "Origin of Species" was published, and without the incorporation of any notion that modern biologists would recognize as evolutionary . . . It is therefore hardly surprising that a younger generation of palaeontologists had so little trouble in assimilating the Darwinian *mechanism* of change into a larger-scale picture that needed little modification in order to be made evolutionary.[34]

Owen's changing philosophy was also signalled at a taxonomic level. Classification does not necessarily reflect 'reality', at times it may serve to legitimise one's philosophy, for example by reshuffling nature into a more compatible form. Looking back again to

1841, Owen needed to fortify his position that reptiles had hit a Mesozoic peak before degenerating. He achieved this by dignifying the fossil "lizards" in question with a title and rank befitting their high station – hence the order Dinosauria, with its anti-Lamarckian overtones. It follows that a different image of nature may require an amendment to the classification, and in 1859 we find a beautiful example of this. As "deeper insight" had convinced Owen that fish and reptiles were artificially divided, so he attempted to impart the idea of continuity by creating a bridging order. Previously the lowest reptiles had been housed in the Labyrinthodontia (containing *Archegosaurus*). Now Owen erected a still-lower order Ganocephala ("lustre-head" after the shiny black skulls common to archegosaurs and lung-fish). Needless to say, Huxley – who had *denied* "geologic" progression throughout the fifties – found the name "inadmissable" because labyrinthodonts were also lustre-headed. Hence he left *Archegosaurus* where it was.[35] But foreigners less caught up in the antagonisms of British science were more sympathetic. Ernst Haeckel in his *History of Creation* was still using Owen's order in 1876, and Owen's American friend E. D. Cope resorted to it as late as 1885. Owen himself admittedly limited inclusion in the Ganocephala to "the so-called *Archegosauri*", arbitrarily excluding *Lepidosiren*. Nevertheless the bridging order was designed to show the fish-like foundations of the entire reptilian class, and as such signalled Owen's acceptance of fossil continuity.

Structurally speaking, Owen's palaeontology of the later 1850s is hard to distinguish from post-Darwinian approaches. He employed diverging pathways and generalised antecedents; he abolished major boundaries and established a pattern onto which the Darwinian 'tree' could be mapped. Finally, Owen's deployment of *Archegosaurus* as an "annectant" form to bridge two classes, suggests that his philosophy had greater heuristic value than Huxley's jibes about his "mystifications" might allow.

Owen's overall picture might actually have been crucial to an early *rapprochement* with an evolutionary palaeontology. At least his doctrines were widely discussed in the 1850s and themselves made the basis of several "derivative" systems. Baden Powell had pressed the archetype into service in his *Essays*, unashamedly putting a transmutatory gloss on Owen's "stately steps". And take the case of Major-General Joseph Portlock (1793–1874) of the Royal Engineers, a stratigrapher who had successfully traced Murchison's

Silurian system into Co. Tyrone. At 64 – and a veteran of the War of 1812 against the United States – Portlock was no sprightly youngster when he assumed the Chair of the Geological Society (1857–8). All the same, his presidency was distinguished for "sound judgment", according to Murchison, and age was evidently no bar to his progressive affiliation. Portlock told the Society in 1858 that Owen's theory of generalised antecedents was "unquestionably one of great importance in speculative Palaeontology", and that the "gradual change" Owen envisaged "from the more general type of vertebrate organization to a more special type" was barely compatible with the idea of "original creation".[36] To Portlock it made more sense as a "progressive modification" during "a long series of ages". So from a purely palaeontological viewpoint, Darwin's *Origin* appeared at an opportune moment, when Owen's mature views on divergence and specialisation were at their most compatible.

Owen and the Darwinians

Enough has been said to explain why Darwin thought himself cruelly deceived at that December meeting. Owen's philosophy of "continuous creation", or uninterrupted origination of life, hooked up to a tentative metagenetic mechanism, mimicked evolution in important respects and led to similar deductions concerning the fossil record. Conversely, Owen's need for an ordered, physiologically-sound explanation of species production which might reflect a "predetermining Will" ruled out Darwin's "higgledy piggledy" selection. The basic stumbling block for providentialists was that the *Origin* was really a misnomer; Darwin had *assumed* the production of novel forms and concentrated on a sifting mechanism to preserve the best. As the Professor of Anatomy at Harvard, Jeffries Wyman, confessed to Owen, "while I admire the genius of the man, & his contribution to science, I cannot avoid the conviction that Mr Darwin has mistaken a distorting element for an efficient cause".[37] Owen's constant auditor, the Duke of Argyll, made the same point in *The Reign of Law*: natural selection explains the "success and establishment and spread of new Forms when they have arisen", but gives no clue to their cause of origin, which must be "by some other law". In other words, as a *Quarterly* reviewer succinctly put it, selection "weeds, and does not plant". As early as 1837, Owen had begun taking a physiological approach to the species problem, and this must partly explain his inability to come to terms with the *Origin*. Because Darwin failed to investigate the generative laws producing a new variant, Owen uncharitably accused him in 1860 of "leaving the determination of the origin of species very nearly where [he] found it" – not exactly what he had intimated the previous December! In print Owen had trouble conceding a single point, and his *Edinburgh* review was disingenuous to the extent that he ignored *Archegosaurus* altogether, surely the single most crucial piece of fossil evidence. His review has variously been dismissed as confused, vindictive, and jaundiced, each with some justification. On the other hand, it is easy to misjudge him from our oblique Darwinian vantage point, especially with him waving the incongruous and apparently irrelevant facts of metagenesis in Darwin's face. But a method lay behind this apparent madness – he was casting about for a generative process that could produce a new species *and* be seen in operation today. His seemingly bizarre and off-the-point remarks on the *Origin* actually reflect long-considered thoughts on the 'species question' and ways to solve it within conventional science.

We can guess why Owen was "atrociously severe" on Huxley in the *Edinburgh Review*. George Rolleston was "much pained by Owen's article" but reasoned to Huxley, "You had some forecast of what was to happen I suppose as you have given him as good . . . in your review."[38] Inevitably passions spilled into the Darwinian arena, explaining Owen's occasional callous remark – but to understand his overkill reaction we must look a little deeper. Darwin was shocked by Owen's "spite" and lost a night's sleep, "The Londoners say he is mad with envy because my book has been talked about", he told his old Cambridge teacher J. S. Henslow. Jealousy is a plausible motive, but it was precisely because of his sense of self-importance that Owen could be so upset by Darwin's talk of the "blindness of preconceived opinion" and of the "ignorance" masquerading under such terms as "plan of creation". And he was moved to anger by Darwin's caricature of Creationism, in which elemental atoms miraculously flashed into living tissues – "preposterous and unworthy" he called this. Accordingly he accused Darwin of misrepresentation and "indifference to the matured thought and expressions of some of those eminent authorities on this supreme question in Biology". Of course Owen had himself in mind, although *The Saturday Review* thought that if he had been ignored in the *Origin* it was his own fault for refusing to step beyond the bounds of dry inductive logic and really argue his case.

Without question Owen was prey to contemporary fears for the status of mankind, and this sensibly influenced his reaction to the *Origin*. Ironically, at the time he was pushing fish and reptiles closer to provide the most compelling instance of pre-Darwinian continuity, he was pulling man and ape apart, showing that moral responsibility and human freedom were altogether higher concerns. Again, he was not denying the possibility of change, actually hinting in 1863 that man might be replaced by "another, probably a higher, specific form" at an inconceivably remote date[39] – only that *transmutation* was the mechanism. Owen was arguably the world's foremost authority on ape osteology, partly as a result of his access to zoo cadavers. But it is probably true that he had been strongly motivated to conquer this field, since his intent was always to remove the threat of bestialisation. For example, his first memoir on the chimpanzee (1835) concentrated on changing muzzle shapes. The baby ape with its deceptively human physiognomy was well known to anatomists, but here Owen was to prove that it acquired a prognathous and rather brutish snout as an adult – a fact to make Lamarck's ape ancestry look decidedly suspect. Again in 1849, following *Vestiges* and the Chartist violence, Owen described the gorilla (newly-discovered by the Episcopalian missionary Thomas Savage in West Africa), while denying that the ape's canines or brow ridges were modifiable characters, or that the formidable brute could transmute into man. In a similar way, Lyell's non-progressionist geology was a reaction to the Lamarckian threat, particularly to human "dignity", according to Bartholomew. But whereas Lyell had rationalised transmutation by the later 1850s (even if he left the fossil pathway preordained), Owen was still casting around for an alternative means of human 'birth'.

The recent arrival of the gorilla undoubtedly explains its novelty in the Darwinian controversy. More importantly, it re-emphasised a problem familiar since the eighteenth century. When Owen published a lecture "On the Gorilla" in 1859, the Master of Trinity at Cambridge, William Whewell, quizzed him on the anatomical explanation of man's superiority to the ape: did the nerves or muscles of the hand or tongue make the difference, "or is the mind working on almost the same anatomy"?[40] This structural similarity was a fearful problem. Owen chose to tackle the problem at the cerebral level, presumably because man was a conscious decision-maker and "especially adapted to become the seat and instrument of a rational and responsible soul". He searched for structural peculiarities in the brain which might explain man's unique behaviour and permit his taxonomic separation. Astonishingly, he erected a human subclass Archencephala ("ruling brain") in 1858, justifying it on the grounds that a lobe, the hippocampus minor, was peculiar to the human brain, and that the cerebral hemispheres completely covered the cerebellum; in apes, he alleged, the hippocampus was absent and the cerebellum was always exposed from above. It is odd that he should have made an error of such magnitude. Huxley blamed it on his cribbing of figures of spirit-hardened and distorted brains, and Lankester on poorly supervised "dissectors and draftsmen". But the fact is that Owen had been working with preserved brains for thirty years, and the leading Dutch specialist, W. Vrolik, had been sending him papers on the "encephalon of the Chimpansé" for checking since the early 1840s.[41]

It is hardly coincidental that Huxley's Royal Institution course on "The Principles of Biology" included a lecture "On the special peculiarities of man" (16 March 1858). Here he lined up man, gorilla and the monkey *Cynocephalus* and pointed out their skeletal and cerebral continuity, introducing the proposition made famous in *Man's Place* five years later:

> Now I am quite sure that if we had these three creatures fossilized or preserved in spirits for comparison and were quite unprejudiced judges we should all at once admit that there is very little ["if any" deleted] greater interval *as animals* between the *Gorilla* & the *Man* than exists between the *Gorilla* & the *Cynocephalus*.[42]

Moreover, Huxley's naturalistic proposition that even "the mental & moral faculties are essentially & fundamentally the same in kind in animals & ourselves" set him determinedly on a collision course

with Owen (as well as powerfully disposing him to an evolutionary
explanation of man's origin eighteen months later). So in 1858
Huxley failed to note any anatomical distinction between the
brains. But Owen's crime was not that he erred. His inability to
retract any point except by way of evasive manoeuvre finally
allowed Huxley to indict him for perjury. Huxley's campaign was
so devastatingly effective because he turned the issue into "one of
personal veracity". The *Natural History Review* in the early 1860s
became a vehicle for anti-Owenian polemics, and he took papers
from Flower, Rolleston, Vrolik, and others, few of whom were
entirely happy with Huxley's "combative" spirit, but were clearly
thrown together by Owen's recalcitrance. Most, being Christians,
held Rolleston's view that psychological differences were anyway
independent of cerebral ones,[43] and that the issue was ultimately
unimportant because "our higher and diviner life is not a mere
result of the abundance of our convolutions". By now it was
apparent that Owen had made a monumental error of judgment.
Not only did his reputation take a needless battering, but he
practically invited Huxley to turn the anti-evolutionary implications on their head. Owen himself had introduced the stakes,
running part of his own campaign under the banner "Ape-Origin of
Man as Tested by the Brain". He evidently thought he could gain
public sympathy by denouncing Huxley as an "advocate of man's
origin from a transmuted ape", but the ruse backfired. That the
debate proved nothing evolutionarily was now beside the point.
Because Owen gambled man's ancestry on the outcome (or so it
seemed), his exposure inevitably turned into a psychological victory for Darwin.

Privately Huxley took steps of a different kind. In November
1862 he was "greatly & unpleasantly astonished" to find Owen's
name up for adoption to the Council of the Royal Society, and in a
letter to its Secretary, William Sharpey, made it plain that this was
unacceptable. Considering

> that one of us two is guilty of wilful & deliberate falsehood, I
> did not expect to find the Council of the Royal Society
> throwing even a feather's weight into the scales against me.
> But of the fact that Owen's selection onto the Council at this
> particular time will be received & used in that light there
> cannot be a doubt.[44]

Sharpey was an able physiologist, immensely respected as a teacher at University College, and quite at home with the cerebral issues. He knew Huxley, having helped him job-hunting on innumerable occasions, but ideologically perhaps he inclined more towards Owenism. For example, in 1862 he still thought the archetypes and homologies of "surpassing beauty and interest", even though a natural explanation (like Darwinism) could only "exalt our conceptions of creative wisdom." On Society matters, he pointed out that it was five years since Owen had sat on the Council (the time interval between Huxley's appearances). As to the brain, "No one", he said, "is more opposed to him on that question than I am", but the Society must refrain from taking sides. Huxley replied that such impartiality was laudable in a "fair controversy". This was no longer a debate over colourless facts but an issue of moral responsibility, and therefore a question of "whether any body of gentlemen should admit within itself a person who can be shown to have reiterated statements which are false & which he must know to be false.' Sharpey clearly had no intention of striking Owen's name off, but he did forward Huxley's letter to the elderly President of the Royal Society, Edward Sabine, who was utterly dismayed, considering it "a *very painful* one" and praying "that we have no occasion to believe that one or other of those in controversy have been 'guilty of wilful & deliberate falsehood'. It would indeed be a painful position for the Society to be obliged to take either side in a *moral* dilemma of so serious a character."

Owen was elected on 1 December 1862, but Huxley's tenacity still needs explaining at a more fundamental level. He evidently intended to create a spectacle, telling Dyster that October: "Before I have done with that mendacious humbug I will nail him out, like a kite to a barn door, an example to all evil doers but it is a slow & wearisome process".[45] Indeed it was, lasting all of three years (1860–3). Part of the problem was Owen's refusal to lie down. But older and more chivalrous onlookers like Sabine still saw Huxley step well beyond the bounds of good taste. Perhaps then we should consider the broader social strategies of the new-breed scientists in the 1860s. For them 'Truth' was a means to power, a legitimisation of their right to professional self-determination and autonomy from the Church. Thus it had to be *seen* to be the final arbiter. A gentleman's agreement not to pursue the issue was no longer any good. Younger biologists seeking professional respect had to adopt a more aggressive course; it was strategically important for Huxley to prove moral laxity, hence he was willing to take the time "to get a lie recognised as such." It would serve his social group well, a group with no sympathy for theology or Platonism, and united in their middle-class hatred of Owen's social climbing.

At a more mundane level it also served Huxley well. The theatrics used in pinning out Owen boosted his book sales. Lyell admitted on reading Huxley's lecture on the "Zoological Relations of Man to the Lower Animals" that "The public will at present devour any amount of your anthropoid ape questions".[46] *Man's Place in Nature*, incorporating this lecture and a blow-by-blow summary of Owen's manoeuvrings, sold its print run of 1000 copies briskly and in March 1863 Huxley informed Dyster "that we are printing our second thousand". By then the debate had acquired an almost international air, with a rush of papers on classification. The anti-materialist Rudolph Wagner came out of semi-retirement and confided to Owen, "Long have I stood on your side", whereupon he "joined issue" with another "discussion on the skull & brain of men and apes" (from which it is clear that he sympathised more with Owen's ideological struggle than his facts). The debate spread rapidly to the colonies, and in Australia, W. C. Thomson passionately defended *Man's Place* against Owen's protégé George Halford, Professor of Anatomy at Melbourne University and author of a rival tome on monkey anatomy engagingly-titled *Not Like Man*.

All this left Owen in an embarrassing position. It also raised the question of exactly how he envisioned the arrival of man, if not by transmutation or direct creation. Alternation of Generations was fine for polyps and medusae, actually supposing the life cycle could be broken, but mooted in connection with the 'birth' of man from an ape it was harder to envisage. Although not impossible – von Kölliker did take this kind of speculation one stage further in 1864 and suggest that new species arose perfectly adapted by a saltatory jump, on analogy with this invertebrate alternation. Much as Huxley disagreed with Darwin over *gradual* transmutation, he still failed to see this as any better solution.[47] Owen too was moving insensibly from metagenesis to saltation, reaching his definitive position in *Anatomy of Vertebrates* (1868). On man, though, he remained tight-lipped, except to throw out some seemingly casual but tantalising hints in his letters. For example, the Rev. Gilbert

Rorison wrote after reading the *Edinburgh Review*, apparently unaware that he was addressing the author, and suggested that the reviewer's position

> seems to be that the higher animals, including Man, were produced by *Creative Law* (& in so far supernaturally) *through* the lower: e.g. the womb of the Ape, or some other high mammal, was the nidus or matrix of the primitive Foetus of Man – this mediation of the lower structure being the mode of a special Creative Energy.[48]

Owen refused to be drawn, simply referring Rorison to his *Palaeontology* and admitting his belief in "a pre-ordained continuously operative secondary cause or law".

> Of the nature & mode of operation of that law I have . . . insufficient evidence. But, should it ultimately be permitted to attain to that knowledge, by experiment & observation I cannot at all answer how it should diminish our sense of responsibility for the use of the powers temporarily entrusted to us, even though 'original sin' should prove to be only the remnant of the untransmuted ape still lingering in our constitution.

The reply is astonishing for Owen's willingness to explain original sin as a result of the malingering ape within; mankind simply represented the definitive modification of the archetype, *superimposed* over an inferior ape pattern by some means other than transmutation. It is also astonishing for being written to an Episcopalian minister from Aberdeen, author of the reactionary "Creative Week" in *Replies to "Essays and Reviews"* (1862), who not only denied transmutation, but castigated any natural explanation as atheistic. For Rorison the geological strata registered "divine acts strictly creative and supernatural" and man came direct from his Maker's hands.[49] Oddly enough, Rorison asked to quote Owen's letter, oblivious to the fact that his sympathies lay with Powell in the offending *Essays and Reviews* – and while failing to see that what he quoted, though eminently pious in tone, was in some ways a gentle chide to himself, viz.:

If at every stage, in the system of our planet, Final Purpose, Divine Adaptation of means to ends, be discernible, and by the Best discerned, do we, in recognising as God's ministers certain secondary causes, necessarily lose sight of Him?

Far from taking a doctrinaire line, Owen was *actively* urging Christians to disregard the literalist teachings of their youth – to beware "of logically precise and definite theologies . . . claiming to be final and all-suffficient." The title of his Exeter Hall talk to the YMCA in November 1863, "Instances of the Power of God as Manifested in his Animal Creation", sounds fearfully orthodox. But this was a radical and passionate attack on sectarian dogmatism, in which he implored young Christians to "Emancipate yourselves from [childhood] notions of textual meaning" and "become again 'as little children'" before the scientific insights into Creation. Not to be frightened by them, not to "look with suspicion, dislike, or dread upon the evidences" as the Cardinal Inquisitors once did upon Galileo's; rather to create a truly charitable Christian atmosphere which would welcome the scientific revelations. The science he had in mind centred on the inductive discovery that fossil creation had been continuous since Silurian times. And Owen took as his text John, Chapter 5, verse 17, in which Jesus said *My Father worketh hitherto, and I work.* "We discern," said Owen of the continuous Presence, "no evidence of pauses or intermission in the creation or coming to be of new species of plants and animals."[50] So strong was his anti-doctrinaire call to arms that the *British Standard* accused him of "covertly thrusting at Revelation". Lyell was quite taken aback and told Huxley:

> He has certainly come out in a new character. The free thinking is nothing new to *us* but that he should treat the parsons so contemptuously . . . is strange & inconsistent when one recollects the line he had previously taken. They will resent it more than all you have done.[51]

The low-brows were predictably outraged and the Darwinians nonplussed. But for those ideologically in line it was different. The new *Geological Magazine* thought that Owen had enunciated "doctrines which the vast majority of scientific men have long accepted", although recognising that "In these days of Büchner and

Vogt" they would no doubt prove "unsatisfactory to the ultra school of developmentalists". They had obvious appeal to anti-Huxleyans like Charles Carter Blake at the Anthropological Society (whose "coarse attack" on *Man's Place* had forced Huxley to resign his Honorary Fellowship in May 1863 and dismiss the society as a bunch of "quacks"). Blake welcomed the Exeter Hall lecture that Christmas, telling Owen that everybody was "delighted with the reverent tone". In particular, "The High Church party are much pleased, as they generally count amongst them the most intelligent and liberally minded people."

On reading proofs of *Man's Place* in August 1862 Lyell was appalled at some paragraphs which were "not in good taste & will do no good", and he advised Huxley "that the best way to carry people with you is to write as if you were not running counter to their old ideas."[52] It was a tall order and, he admitted, easier "to preach than to practise". Indeed he was up against not only Huxley, but the militant strategy of the younger agnostics. Huxley of course was notoriously uncompromising. "Theology & Parsondom", he told Dyster in 1859, were "the natural & irreconcilable enemies of Science". In February 1860 his lectures at the Royal Institution were attended by a bishop and a dean; "to please *them*", he said, he wound up with the most confrontationist reappraisal "of Science *versus* Parsonism that is ever likely to have reached their ears." Partly, this gladiatorial attitude stemmed from the restructuring of the social base of science in the industrial age, and from the need felt by its new middle-class leaders for professional autonomy. In trying to distance themselves from the Church, they actively fostered the 'warfare' image. But Owen was a generation older, and his strategy, like Lyell's, was a more cautious attempt to carry conservatives with him – the very conservatives Huxley would have been happy to see locked out. In light of this, one wonders whether Owen really was "priming" Bishop Wilberforce for the famous Oxford debate in 1860. Enough has been said to put forward an entirely plausible alternative, i.e. that he was helping Soapy Sam see his way to an enlightened view of creation. Not evolutionary, certainly, but then not reactionary. It makes some sense, given Owen's understanding of "continuous creation" and his subordinate speculations. Wilberforce, slated by *The Daily Telegraph* as another of those "old–style Tories" who "have not advanced one iota beyond their ancient notions",[53] apparently needed coaxing into the modern world, and as Roy MacLeod says, Owen can have had little sympathy for the bishop's Mosaic view of creation. Whatever the truth, it turned out to be another tactical blunder, and Owen suffered by association when Wilberforce resorted to "round-mouthed, oily, special pleading".

For all of Huxley's needling, Owen's professional disgrace was largely his own fault. And the fact remains that he was not able to meet the Darwinian challenge. His *Palaeontology*, published in April 1860 four months after the *Origin*, was thrust into a world looking for larger answers. The *Athenaeum* greeted it with a yawn, looked to its subtitle, *A Systematic Summary of Extinct Animals*, and warned "Let no reader . . . expect more from this volume than what it professes to give". Factually, it became a standard reference tome, still in use a generation later. But the press gave it a rough ride because they were looking for a different book:

> We entirely agree with his observation, that "perhaps the most important and significant result of palaeontological research has been the establishment of the axiom of *the continuous operation of the ordained becoming of living things*." And this is the very topic on which we should desire to have his disquisition.[54]

"Mr Darwin has written largely and boldly in the opposite direction, and has found a bold and able advocate in Prof. Huxley. Is Prof. Owen, then, to be reserved?" What was needed was an authoritative statement on his philosophy of "continuous creation".

In the second edition of *Palaeontology* (1861), Owen quietly rewrote his conclusions on *Archegosaurus*, the "annectant" amphibian so conspicuously missing from his *Edinburgh* review. He pointed out that to anyone searching for the origin of species, this was "the most exemplary instance of a transitional form, on the derivative hypothesis, of an air-breather from a water-breather." Precisely how "one form, or grade of animal structure may be changed into another", say a *Lepidosiren*-like fish into an armoured amphibian, was an immense problem. Whether the cause was "external and impressive, or internal and genetic", or both, was the stuff of future research. One thing was certain: that however inadequate Lamarckism and Darwinism, the possibility of such "derivation" *was* still "open to the mind of every unbiased explorer of the laws of animated nature; and no fact, old or new, ought to be dismissed until its relations to the great question have been completely and impartially considered . . ."[55]

Like his auditors, St George Mivart and the Duke of Argyll, Owen rejected the *Origin* for its ugly mechanism; unfortunately a newly industrial age whose aesthetics were radically changing failed to share his vision of beauty. Even had he written a first-rate account of "continuous creation", as the *Athenaeum* wanted, it would have been too late. Nor did he really suppose it necessary, imagining that Darwinism would collapse spontaneously, and a more empirical and plausible science rise to take its place. Such was not to be the case – the social and ideological foundations of science were being transformed at an alarming rate, and Industrial society, based on competition, *laissez faire*, and Spencerian principles, found its interests well served by the *Origin*.

3

Huxley's 'Persistence'

. . . I confess it is as possible for me to believe in the direct creation of each separate form as to adopt the supposition that mammals, birds, and reptiles had no existence before the Triassic epoch. Conceive that Australia was peopled by kangaroos and emus springing up ready-made from her soil, and you will have performed a feat of imagination not greater than that requisite for the supposition that the marsupials and great birds of the Trias had no Palaeozoic ancestors belonging to the same classes as themselves. The course of the world's history before the Trias must have been strangely different from that which it has taken since, if some of us do not live to see the fossil remains of a Silurian mammal.

Huxley being provocative in 1869.[1]

The Oxford Darwinian E. B. Poulton long ago observed that Huxley's presuppositions often radically differed from Darwin's; and we know that his vindication of the *Origin* was not always what Darwin would have wished. Yet curiously little attention has been paid to Huxley's pre-Darwinian position – with the exception of his saltatory views (his belief that "*transmutation* may take place without transition" as he suggested to Lyell in 1859[2]). Almost nothing has been said of his non-Darwinian biogeography, and only now are we beginning to tackle his anachronistic attitude towards classification and the fossil record.

In a recent revisionist statement, Michael Bartholomew highlighted the incongruities of Huxley's defence of Darwin. Here I want to concentrate on Huxley's *geological* understanding, and to consider the strengths and weaknesses of his objection to the most sophisticated form of progressionism – the idea of progressive divergence and specialisation. In the later 1850s he took an uncompromising stand on what he called "persistent types" – fossils which remained virtually unchanged through immense periods of geological time. In fact, his theory in its strongest form – say in 1859 – indicated that a perfect fossil record would reveal *no* progressive trends whatever, and even if development had occurred in "pre-geologic" times, we could have no direct evidence of it. Because progressive divergence was so essential, not only to Carpenter and Owen, but also to Spencer and Darwin, Huxley's intransigence led to a certain conflict of loyalties. His time-warped perception of the emergence of fossil life put him in a rather odd position when donning his Darwinian hat, and the 1860s and 1870s saw him trying to square his geological views with the new Darwinian evolution. In this chapter we will also look at the way his theory of 'persistence' was related to other non-Darwinian aspects of his thought – it may, for example, explain why he so eagerly summoned up a lost continent and used it as a jumping-off stage for the diffusion of fossil life. But most of all, when we come to look at his attempt at ancestor-tracing, it should help to explain why he could happily handle only small-scale 'phylogenies' (particularly horses and crocodiles, i.e. developments within the family), but was severely handicapped when faced with the emergence of new *classes*. Indeed, he had to adopt drastic *ad hoc* measures to explain why the bones of mammals and birds did *not* turn up in Silurian rocks, but only much later in Mesozoic deposits.

Huxley was possibly the only post-Darwinian palaeontologist who could envisage mammals in the Silurian; never an empirical literalist, it mattered little to him that sediments from this period preserved nothing higher than molluscs and a few fishes. He even speculated that man himself might be a "persistent type". Closing *Man's Place in Nature* (1863), he asked: "Where, then, must we look for primaeval Man? Was the oldest *Homo sapiens* pliocene or miocene, or yet more ancient? In still older strata do the fossilized bones of an ape more anthropoid, or a Man more pithecoid, than any yet known await the researches of some unborn palaeontologist?"[3] Even this was a greater concession than he was prepared to make in 1859. Man was possibly so old, he speculated in a letter to Lyell, that we might never find his monkey-like ancestors in the rocks. And to round off, he vividly pictured *Homo ooliticus*, a hypothetical dinosaur contemporary, clubbing the fossil 'opossum' *Phascolotherium* "as the Australian niggers treat their cogeners".

Setting the Scene – Huxley's Use of Amphibians

> . . . I cannot help hoping that you are not quite as right as you seem to be.
>
> Darwin gently remonstrating with his disciple.

Shortly after publication of the *Origin*, and while Owen tackled *Archegosaurus* and its 'ancestry', Huxley himself started serious work on the labyrinthodont amphibians. But curiously, although he sharpened his beak and claws ready to defend Darwin, he failed to use the new fossils to prove descent; in fact, their significance for him lay elsewhere. Throughout the sixties he subjected them to meticulous study, but his dry descriptions left not the slightest hint that he was an evolutionist, let alone a Darwinian. Some of his new fossils were later to be of critical importance in deciding reptile ancestry – a complete specimen of *Pholidogaster*, for example, which the knightly team of Sir Philip Egerton and the Earl of Enniskillen had placed in the British Museum, was named and described by Huxley in mid-1862. Later that year he introduced the advanced anthracosaurs, amphibians from the coalfields of Airdrie, in Scotland. But so far from volunteering thoughts on affinities, he was strangely reluctant to do more than name and describe – to the extent that one referee finally pulled him up and requested information on the "systematic relations and affinities" of these animals.[4] Even then Huxley scrupulously avoided any evolutionary implications; nor did he mention Owen's archegosaur-lungfish lineage – although he took time out to obliterate his new order Ganocephala. Huxley's heart lay elsewhere, as we find out when we consider the use to which he eventually put these amphibians. *Anthracosaurus* with its sturdily built spine, persuaded him not that the group had undergone any progression, but the reverse: that it was one of the most remarkable examples of static existence on record. So many Carboniferous amphibians – some as highly developed as their Triassic kin – had turned up by 1870 that he was able to hold them

Era	Strata	Period	First recorded appearances of fossil animals	Faunal group
TERTIARY or CÆNOZOIC	Turbary. Shell-Marl. Glacial Drift. Brick Earth. (Bone-Caves.)	Pleistocene	MAN by Remains. ... by Weapons.	Birds and Mammals.
	Norwich, Red, Coralline } Crag.	Pliocene		
	Faluns. Molasse.	Miocene	Ruminantia. Quadrumana. Proboscidia. (BIRDS, Orders of.) (MAMMALS, Orders of.)	
	Gyps. London } Plastic } Clays.	Eocene	Rodentia. Ungulata. Carnivora.	
SECONDARY or MEZOZOIC	Maestricht. Upper Chalk. Lower Chalk. Upper Greensand. Lower Greensand.	Cretaceous	Cycloid } Ctenoid } FISHES. Mosasaurus. Polyptychodon. BIRDS, by Bones. Procœlian Crocodilia.	Reptiles.
	Weald Clay. Hastings Sand. Purbeck Beds. Kimmeridgian. Oxfordian. Kellovian. Forest Marble. Bath-Stone. Stonesfield Slate. Great Oolite. Lias. Bone Bed.	Wealden / Oolite (L. M. U.)	Iguanodon. Marsupials, — Chelonia by Bones. Pliosaurus. Marsupials. Icthyopterygia. MAMMALIA. (Amphicœlian Crocodilia, Pterosauria, Homœocercal Fishes, Cephalopods 2-gilled.)	
	U. New Red Sandstone. Muschelkalk. Bunter.	Trias	AVES, by Foot-prints. Sauropterygia. Labyrinthodontia. (Crustacea 10-poda.)	
PRIMARY or PALEOZOIC	Marl-Sand. Magnesian Limestone. L. New Red Sandstone.	Permian	Sauria. Chelonia, by Foot-prints. (Isopoda.)	Fishes.
	Coal-Measures. Mountain Limestone. Carboniferous Slate.	Carboniferous	REPTILIA ganoceph. Insecta.	
	U. Old Red Sandstone. Caithness Flags. L. Old Red Sandstone. Ludlow.	Devonian	PISCES { ganoid. placo-ganoid. placoid. (Heterocercal.)	Invertebrates.
	Wenlock. Caradoc. Llandeilo. Lingula Flags. Cambrian.	Silurian	Echinoderms. Annelids. Bivalves. Trilobites. Pteropods. Brachiopods. Gastropods. Cephalopods 4-gilled. Fucoids. Zoophytes.	

Table of geological strata and first recorded appearances of fossil animals. (From Owen [1861a].)

aloft as prime examples of 'persistence', in this case through a prodigious lapse of time "represented by the vast deposits which constitute the Carboniferous, the Permian, and the Triassic formations."[5]

It is revealing that "Darwin's Bulldog" should have concentrated on evidence which, although technically neutral (and which in the *American Addresses* he calls "neutral" or "indifferent"), was nonetheless generally considered unfavourable to Darwin. Huxley was aware of the delicacy of the situation. After his address on 'persistence' to the Geological Society in 1862 he dropped Darwin a note to allay fears that he had deserted the cause:

> I want you to chuckle with me over the notion I find a great many people entertain – that the address is dead against your views. The fact being, as they will by and by wake up [to] see that yours is the only hypothesis which is not negatived by the facts, – one of its great merits being that it allows not only of indefinite standing still, but of indefinite retrogression.

The last thing Darwin wanted to do was chuckle at this strange exhibition. It was not that Huxley was wrong; indeed, it was imperative that someone pointed out that only the *Origin* could cope with a fossil record of fits and starts. But his presentation was so atrociously one-sided, with almost all the emphasis on 'persistence' to the total exclusion of progression. This more than anything had Darwin worried, and he wrote back courteously:

> I can say nothing against your side, but I have an "inner consciousness" (a highly philosophical style of arguing!) that something could be said against you; for I cannot help hoping that you are not quite as right as you seem to be. Finally, I cannot tell why, but when I finished your Address I felt convinced that many would infer that you were dead against change of species, but I clearly saw that you were not. I am not very well, so good-night, and excuse this horrid letter.[6]

Darwin would have liked a little more progression and a little less 'persistence'. His pugnacious young disciple, though willing to risk the stake for the cause, was unable to accommodate.

Huxley's dogged adherence to 'persistence' helps explain his palaeontological priorities, in particular his reluctance to speculate on the origin of major groups. Take the amphibia – Owen had pictured them arising from something like *Lepidosiren*; but this was a *living* lung-fish, and it was rapidly becoming apparent in the wake of the *Origin* that actual ancestors would have to be sought in the appropriate rocks.[7] The trouble was they could so rarely be found, and this led to some ingenious "master wrigglings", as Darwin would have called them. In Jena, the crusading Darwinian Ernst Haeckel got over the problem by resorting to hypothetical "ante" periods, placed before each period but subsequently wiped from the record (being a time of elevation and therefore weathering). Thus the Triassic reptiles all evolved in ante-Triassic time, and presumably the archegosaur's ancestors lived in the ante-Carboniferous, and their fossils were subsequently destroyed. "Wholly incredible", was Huxley's predictable response. Though he sympathised with Haeckel's monistic materialism (even if it proved a little too severe for his tastes), he disagreed "fundamentally and entirely" with Haeckel's timing of geological events. He even accused him of being "overshadowed by geological superstitions".[8] Huxley's arguments against some of the "ante" periods were decisive. For example, the Permian and Triassic deposits could actually be seen running into one another, squeezing out any ante-Triassic. But the real reason for Huxley's sudden descent on his brother *idéologue* was Haeckel's standard chronology. He was happy to have each group evolve just prior to its first appearance in the fossil record; in other words, the rocks – for all their preservational bias – must be our ultimate guide. Huxley had grave misgivings. He might have asked why, if Carboniferous labyrinthodonts could persist into Triassic times "with no important modification", they could not have come down from Cambrian times unchanged? Likewise, why should "the development of Mammals, of Birds, and of the highest forms of Reptiles" be crowded into a conjectural ante-Trias? More likely, they had persisted for untold ages, and he was prepared to put their origin "far back in the Palaeozoic epoch".

However implausible Huxley found "ante" periods, Silurian mammals required a greater leap of faith. But so long as he was prepared to make this leap, any talk of amphibian origins might have to be shelved. After all, where was one to look for evidence? If mammals were already flourishing in the Silurian, amphibians must be inestimably older, perhaps older than the records themselves. So the very exigencies of a theory of non-progression, in effect, *forced* Huxley to steer clear of detailed discussions of amphibian origins.[9]

Why 'Persistence'?

It was originally assumed, for example by the historian of ideas Arthur O. Lovejoy, that by harping on the permanence of species,

Huxley was trying to kick away the Cuvierian crutch supporting "the special-creation hypothesis" – in other words, the long-term stability of molluscs, crocodiles and marsupials proved once and for all that the cataclysmic destructions and recreations of life were a figment. But London geologists had long dispensed with catastrophist views and there is no evidence that Huxley was trying to set up so ancient a straw man. True, he did tell Hooker in 1857, "I am going to take Cuvier's crack case of the 'Possum from Montmartre' as an illustration of *my* view",[10] but his address to the Geological Society in 1862 shows what he really had in mind. Noting that most Mesozoic 'marsupials' appeared to be highly developed (on a par with the living opossum or Tasmanian devil), he asked "in what circumstance is the *Phascolotherium* [the Jurassic 'opossum' discovered by Justice Broderip and diagnosed by Cuvier] more embryonic, or of a more generalized type, than the modern Opossum"? In none whatever was his answer, phrased in such a way that he could only be rebutting Owen's palaeontological doctrines, which called for more generalised, embryonic, or archetypal forerunners in the fossil record. So we need not look so far afield to understand Huxley's ploy – in fact it was the kind of response we might have expected to Owen's realist version of progressive development.

Perhaps the man most pleased by Huxley's 1862 address was the erstwhile non-progressionist Charles Lyell (1797–1875), and his reaction and relationship with Huxley reinforces our interpretation. Although Lyell recognised that Huxley had overstated his case, mentioning "many private protests against some of his bold conclusions", he nonetheless called the address a "brilliant critical discourse" which proved that "the progressive development system has been pushed too far" and "how little can be said in favour of Owen's more generalised types".[11]

From the mid-1850s, Lyell himself had been desperately trying to accommodate to a Lamarckian progression, and the traumatic details of his inner conflict were revealed in the private *Species Journals*, edited and published by Leonard Wilson in 1970. Lyell began the 1850s as a vehement anti-progressionist, using the Geological Society platform in 1851 to campaign against developmental views. Again in 1852 he cited anomalous observations, like the supposed tortoise tracks in Canadian Silurian rocks, and Mantell's allegedly Devonian *Telerpeton*, to prove how easily the so-called progressive sequence could be overturned. But by 1856 it was obvious that he was fighting a losing battle, and that modern palaeontology was "drifting towards the Lamarckian theory".[12] The *Species Journals* record his painful adjustment to a progressionist and Lamarckian world-view; and he began rationalising the birth of Man from ape parents (perhaps from Lartet's "oak ape" *Dryopithecus*[13]), while agonising over the loss of human dignity. He tried to envisage man "stealing quietly into the world", sired by a lowly ape – he reasoned to himself – as a Newton is born of ordinary mortals. Yet he could never quite ignore that decisive shattering instant when the first human acquired a soul and escaped nature's bondage.[14]

Lyell was tortured by indecision and never went as far or as fast as Darwin would have liked. I doubt that he ever really expunged all opposition to progression in the early 1860s, for he seemed just as thrilled in 1862 that Huxley could make out such a devastating case for 'persistence'. Indeed, where Huxley identified a counterexample – an advance from the sinuous *Archegosaurus* to the flattened Triassic labyrinthodonts with their well-formed vertebrae – Lyell wrote to remind him that the young American palaeontologist O. C. Marsh had supposedly found a Carboniferous ichthyosaur (which would have made this marine reptile as old as *Archegosaurus*), hoping to dash this remaining case of progression.[15]

Bartholomew may be right in claiming that what first attracted Huxley to Lyell's *Principles of Geology* was its air of uncontaminated naturalism, and that he recognised in Lyell a kindred spirit trying to "free the science from Moses".[16] Huxley of course refused to let the inner scientific sanctum, or its positivist host, be defiled by the crasser sorts of metaphysics, and he abhorred Owen's "mystifications" as much as Judaical superstition. Yet there was a certain irony in his turning to Lyell. The non-progressionist *Principles* might have looked ideologically sound to Huxley, but according to Bartholomew it stemmed from a religious commitment to the "nobility" of man; in fact, it could easily have been born out of repugnance of Grantism and its brutalising effects (remember that Robert Grant was advocating transmutation in his "Fossil Zoology" course not a mile away while Lyell was teaching at King's). So in one sense Lyell's science in the 1830s was as 'reactionary' as Owen's. Owen we know abominated Grantism and carefully circumvented it by adopting an anglicised version of *Naturphilosophie*, which left him holding progressionist views. Nonetheless, Lyell was said secretly to have admired Owen's new subclass for Man, and in his *Journals* toyed time and again with Plato's "Ideas".[17] Despite this, Lyell's scientific statements were carefully wrapped in naturalistic garb, and it is not surprising that, repelled by Owen's idealism, Huxley should have swung into Lyell's camp – with a result that as Lyell at last began reconciling himself to progressive development, Huxley started denying that there was a shred of evidence for it.

This meant that Huxley was forced to take issue with those who were not transcendentalists but who shared Owen's progressionist views, in particular the Unitarian William Benjamin Carpenter (1813–1885). We have already touched on Carpenter's generosity towards Huxley when embarking on his scientific career. He was one of the few to support Huxley on "Parthenogenesis" in 1851; he helped him gain an FRS the following year, and he was invited to Huxley's wedding in 1855. But Carpenter's palaeontology was an almost exact replica of Owen's – to the extent that they crossed swords in 1851 over who first applied the embryological concept of progressive divergence to the fossil record. Although Carpenter later lost any illusions he might have had about Owen, and in the 1850s egged Huxley on at every opportunity, a decade earlier he had actually admired Owen and even aspired to be his "aide".[18] In 1842, while still living in Bristol, Carpenter began a lively correspondence, requesting testimonials, and sounding Owen out (sometimes coming to town for the purpose) before publishing himself. Wanting to take up comparative anatomy professionally, but anticipating difficulties in providing "my family with the bare comforts of life", he moved to London in 1844, still hoping to be found "worthy to co-operate" with Owen at the College of Surgeons.

In the *Principles of Physiology* (1851), Carpenter first applied Karl Ernst von Baer's embryological ideas to the fossil record (as Owen had done in his critique of Lyell that year). For von Baer foetal development was characterised by a progressive specialisation away from the undifferentiated germ; so although, say, all vertebrates start from a common archetype, each individual *diverges* during its development towards its unique adult form. Applied to the fossil record, it suggested a progression from the ancient generalised archetype towards more modern specialised forms. Foetal and fossil divergence thus seemed to be parts of a wider developmental *plan*, and Carpenter considered this a "more satisfying" explanation than that offered in the *Vestiges*.[19] His generalised starting point, the archetype, was probably less a Platonic ideal than a model or extrapolation, in Huxley's sense. However, in the 1854 edition of the *Principles*, Carpenter bolstered his position by taking examples of fossil progression from Owen's works. He noted that "in many of the earlier Mammals, a closer conformity to 'archetypal generality' is to be discovered", as in the case of the three-toed Eocene *Palaeotherium*, which he connected to today's horse by a series of fossils showing progressive toe reduction.

Huxley reviewed the *Principles* in the *Westminster* for 1855. At the outset he was complimentary, claiming for it a "leading place among the most advanced works on the subject" and defending Carpenter against the charge of being "simply a 'compiler'".[20] But not surprisingly Huxley had "a crow to pick" and spent the greater part of the review refuting Carpenter's belief that fossil life was a progression, "not so much from the lower to the higher forms, as from the more general to the more special" and that many extinct animals were thus transitional. Now, fossil divergence of this kind was comforting to Darwin, and Carpenter later hinted, with some justification, that the von Baerian emphasis had "in some degree prepared the way" for the *Origin*. So Darwin had seriously to consider Huxley's objections, to understand the sort of opposition he could expect. Shortly before his death, Dov Ospovat found some notes scribbled by Darwin after his famous meeting with Huxley, Hooker and the entomologist Thomas Wollaston in April 1856. Darwin as we know had invited them to Down for the weekend to sound out their views on evolution. Huxley duly rehashed his criticisms of Carpenter, convincing Darwin that his theory *could* withstand them successfully, and two weeks later he began his "big book" *Natural Selection*.

So the onus was on Huxley to accommodate after 1859, when the *Origin* appeared. Ideologically he sympathised with Darwin, who had plainly risen above the Vestigian's "metaphysic stage" in proposing a naturalistic mechanism for the production of new species. But Huxley was in a quandary. Darwin's palaeontology came closer to Owen's than to Lyell's, with the now-familiar family tree superimposed over an older progressive divergence. However, Darwin's denial that progress was inevitable according to his system provided Huxley's leverage point and finally allowed him to dovetail the two geologies. Surprisingly, this dovetailing was a none-too-messy affair, and Huxley juggled Lyell's lack of progression with Darwin's evolution expertly. What he did was to develop an idiosyncratic half-way house. In 1859 he endorsed the Lyellian doctrine of non-progression, but restricted it to *geological* time (i.e. the period covered by the stratified rocks). He evidently assumed that because "physical forces" have changed little during the geological period, life today is largely what it was in Palaeozoic times (an old Lyellian *non sequitur*; most geologists admitted that continually acting forces of similar intensity could and did work progressive changes in the landscape, which in turn called for new and higher creations). Huxley then reconciled this with Darwinism by further assuming that all major evolutionary steps had taken place in *pre*-geologic time, presumably while the world was cooling and subject to more turbulent forces. As a result, nothing (or very little) new had appeared under the sun since the Cambrian or Silurian period. This explains the parting shot of Huxley's 1859 lecture to the Royal Institution "On the Persistent Types of Animal Life": "all we know of the conditions in our world during geological time, is but the last term of a vast and, so far as our present knowledge reaches, unrecorded progression."[21]

This position was probably too far-fetched to attract any disciples. But we must tread with extreme care because a number of palaeontologists did talk of 'persistence' in the early 1860s. However, they clearly meant something quite different by it, and may have had an almost opposite intent. For instance, Hugh Falconer studied the 'persistence' of elephants and mammoths through Glacial and post-Glacial times specifically to show the inadequacy of "'Natural Selection', or a process of variation from external influences".[22] Falconer agreed with Darwin that mammoths were "the modified descendants of earlier progenitors", but since they ranged so widely, suffered such extremes of climate, yet changed so little, he assumed that some "deeper seated and innate principle" governed evolution. In the same way, Martin Duncan looked at fossil echinoderms and inferred from their occasional 'persistence' across time and space some "inherent power of variation which exists more or less in all animated beings irrespectively of the influence of external physical conditions". In other words, they – like so many contemporaries – were trying to subordinate Darwinian selection, and establish an inner driving mechanism divorced from environmental conditions. Huxley, on the other hand, was using long-term geological stasis to combat the realist and progressionist views of Owen and Chambers, who actually depended on some such inner drive.

Huxley, Spencer, & Owen

The ironies of Herbert Spencer's triangular relationship with Huxley and Owen re-emphasise the conflict of interests caused by Huxley's 'persistence'. By the 1850s the practical distinctions between Huxley's and Owen's palaeontology were as clear cut as their ideological differences. Huxley's faith in Lyellian naturalism had led to his denial of all but the most minor progressive trends in "geologic" time. In stark contrast, Owen had used the Tory *Quarterly Review* to criticise Lyell's non-progressionist and ultimately unChristian geology, and had encouraged the search for examples of vertebrate specialisation. Of course, it was because Owen's ancient prototype was actually an *Ideal* form which underpinned a Platonic world view that Huxley was so antagonistic, putting friends like Carpenter on the spot. Many were adamant in their rejection of an Ideal Archetype and none had much sympathy for Owen's Platonism – yet the majority did have a compelling need for fossil progression.

Huxley's friend and founder-member of the *x* Club, the social philosopher Herbert Spencer (1820–1903), illustrates this clash of interests in graphic detail. Spencer found Owen's and Carpenter's palaeontology compelling because he wanted to use von Baer's "Law" of foetal growth and differentiation to demonstrate the inevitability of social and industrial change. In "Progress: Its Law and Cause" (1857), published in the *Westminster Review*, he drew distinct biological parallels with history, showing society complexifying in an 'embryonic' way. The analogy reached cosmic proportions – from nebulous vapours to moral perfectability, progress was everywhere and inexorable. Or, as Spencer said, von Baer's embryological law

> is the law of all progress. Whether it be in the development of the Earth, in the development of Life upon its surface, in the development of Society, of Government, of Manufactures, of Commerce, of Language, Literature, Science, Art, this

same evolution of the simple into the complex, through a
process of continuous differentiation, holds throughout.
From the earliest traceable cosmical changes down to the
latest results of civilization, we shall find that the transformation of the homogeneous into the heterogeneous, is that in
which Progress essentially consists.[23]

Spencer was a man of large and curious contrasts: an arch
capitalist who steadfastly refused to dabble in the stockmarket; a
staunch critic of poor relief who, in financial distress, accepted
$7,000, collected and invested for him by American sympathisers
led by E. L. Youmans. (This enabled him to complete the *Synthetic
Philosophy*, a monolithic series to which he so single-mindedly
devoted his life that it amounted to an act of "self-enslavement".)
Spencer had a vast and pressing need for progress which developed
out of the political aspirations of his Dissenting youth. He was
Derby-born into a line of Wesleyan radicals. Derby was a town
dominated by Methodist manufacturers and shopkeepers, whose
distrust of the Established Church and its Tory spokesmen led to
widespread radicalism and agitation for political reform. Spencer's
laissez faire and call for a reduction in government interference
(which favoured Established interests and handicapped nonconfor-
mists) clearly had Dissenting roots, and his early demands for social
change sprang from the same cultural source. Though retaining the
moral and political presuppositions of his Dissenting stock, he had
become a free-thinker by the forties, and when he wrote *Social
Statics* (1850) this Dissenting urge for political reform had blos-
somed into a wider philosophy of progress justified by organic
analogies. "Instead of civilization being artificial," he stressed that it
was "a part of nature; all of a piece with the development of the
embryo or the unfolding of a flower."[24] Social reform was under-
written by nature: progress "is not an accident, but a necessity."

Spencer's biographer, J. D. Y. Peel, points to the moral dimension in this need for social progress. He comments that Spencer
came to evolution early (the 1840s) because he started off not "from
a phenomenon to be explained, but from ethical and metaphysical
positions to be established"[25] – namely that man was destined to
reach a more morally-perfect state, and that social and industrial
progress were a means to this end. Spencer needed to establish a

directional change in history – a forward-moving, *progressive*, ceaseless change; and it was this social imagery that was largely back-projected onto nature. Hence his adoption of Lamarckism, which he first encountered in Lyell's *Principles* (proving that Lyell had not been entirely successful in discrediting the Frenchman). With acquired characters inherited, it was possible to overarch from biological to social evolution, bringing both organic and cultural development under one roof. Spencer began casting around for laws which could cover in one gigantic sweep the entire realm of cosmic, organic and social history; a "total system" Peel calls it; one which could give a nation of "self-taught artisans" the proof it needed that social betterment was truly cosmic in its inevitability.

In 1851 Spencer went to hear Owen lecture, thinking it might bear "on the development hypothesis". Presumably he was looking for palaeontological proofs and, as always, he seemed to absorb ambient facts "through the pores of his skin", as the corpulent American philosopher and cosmic theist John Fiske joked ("at least he never seemed to read books", which was also Huxley's impression). Spencer got to know Owen well enough to dine with him. Privately he thought his man–ape dichotomy "anything but logical", and took an instant dislike to the Archetype. On this touchy subject, he ended Owen's 1851 lecture course in "complete disbelief". But he did nothing until reassured by Huxley's Croonian lecture, in which Owen's vertebral skull was ceremoniously dismantled. That was delivered to the Royal Society on 17 June 1858 and three weeks later Spencer was busy "with the onslaught on Owen", preparing a paper of his own for the *Medico-Chirurgical Review*. (Not without some collusion; Spencer leant heavily on Huxley, borrowing his edition of Owen's *Recherches sur l'Archétype* for the purpose.) He told his father the *Archetype* was "terrible bosh – far worse than I had thought." Adding, with an appalling presumption, bearing in mind that Huxley had done most of the donkey work, that he would "make a tremendous smash of it, and lay the foundations of a true theory on its ruins." The article too had its gall. Spencer apologised that he had "little previous acquaintance with the facts", and those he possessed were largely Owen's. Yet on philosophic grounds he felt justified in rejecting the Archetype, or rather Owen's demonstration of his Platonic conclusions, as "manifestly insufficient." It appeared in print long before the Croonian, and Huxley found it "rayther strong, but quite just." What Spencer related to his father was a mite more extravagant:

> Huxley tells me that the article on Owen has created a sensation. He has had many questions put to him respecting the authorship – being himself suspected by some. The general opinion was that it was a settler.[26]

It looks as if Spencer had reached common ground with Huxley, but probably only in his secular and anti-Platonic aims. Here we must be careful because much of the Owenism that Spencer absorbed through his pores *was* immensely comforting to him. Indeed, like Owen a few years before, he had begun his marvellously-titled "Illogical Geology" in 1859 as an attack on Lyell, which ought not to have endeared him to Huxley. As a free-thinker, he might have hated the Archetype, yet the von Baerian nub of Owen's palaeontology – the progressive specialisation – was precisely what characterised Spencer's social views. He revelled in Carpenter's *Principles*, which he found "considerably more useful and vastly more entertaining" than books about "fights and dispatches and protocols".[27] And this was largely because of its von Baerian emphasis; indeed, it was while reviewing the third edition in 1851 that Spencer first came across von Baer, whose embryological views he subsequently canonised as a "Law" in his "Progress: Its Law and Cause". So in his *Medico-Chirurgical* criticisms of Owen the following year, he could hardly take exception to fossil specialisation; quite the reverse, he badly needed it to support his social analysis. What he did do was to explain it by naturalistic Lamarckian means, arguing that "all higher vertebrate forms" arose by "the *superposing of adaptations upon adaptations*" and that species

> have resulted by gradual differentiation under the influence of changed conditions [from which] it would manifestly follow that the higher heterogeneous forms would bear traces of the lower and more homogeneous forms from which they were evolved.[28]

But plainly the image evoked was Owenian (or von Baerian), and Spencer carefully wove Owen's fossil illustrations into his social narrative. Thus Owenism *minus* the Archetype was carried by a wider social philosophy. Hence the incongruity of Spencer's position, his split loyalties; publicly desecrating Owen's Holy of Holies, the archetypal host, yet supporting its apparent fossil implications at a deeper level. An even greater irony was Spencer's applause for Huxley, whose 'persistence' was so clearly inimical to his own radical ethic. One can actually measure Huxley's growing impact by his destructive encroachments on Spencer's unitary system.

Huxley first met Spencer one day in 1852, calling at his *Economist* office to thank him for an offprint. And so began a powerful friendship which, Spencer admitted in his *Autobiography*, "became an important factor in my life." In 1856 he moved to lodgings not five minutes walk from Huxley's house in St John's Wood. Thereafter they had a standing engagement on Sunday afternoons to stroll together up to Hampstead Heath or along the Finchley Road. This they would do in rapt conversation, often on evolutionary topics, with Huxley invariably naysaying each of Spencer's arguments. On the whole, though, Huxley found some "grand ideas" in *Principles of Psychology* (1855), crowed over the *Medico-Chirurgical* "settler", and found little to criticise so long as Spencer remained the iconoclast. "I value his approval more than that of any one", Spencer once said, "as he is always so critical and sceptical, and so chary of his praise." As well he might have been: progress and evolution were prickly points, and Huxley was not averse to shooting Spencer's high-flown ideas in mid-air. He considered himself a "devil's advocate" to Spencer, responsible for choking no end of "brilliant speculations" in "an embryonic state". Reading this admission years later, in Huxley's *Life*, Spencer dismissed it as another of his "facetious exaggerations"[29] (making sure Leonard Huxley printed what amounted to a retraction in *The Athenaeum*). Yet it is not hard to see Huxley and Spencer at odds over fundamental issues, especially before Huxley adopted evolution and availed himself of a modicum of post-Cambrian progress.

On biological points Spencer bowed to Huxley's superior knowledge and made increasing concessions. The *Westminster* "Progress" – written in 1857 while attending Huxley's Royal Institution lectures – was still optimistic and palaeontologically progressive. Spencer cites Owen "as decisive" on the departure of fossil vertebrates "from archetypal generality", and quotes Carpenter to the same effect. Yet he does note that new fossils could be expected to push origins back untold distances, and – in peculiarly Huxleyan vein – that "for aught we know to the contrary" the geologic period may only represent "the last few chapters of the Earth's biological history". However the onus remained on progression, for he concluded:

> we cannot but think that, scanty as they are, the facts, viewed in their *ensemble*, tend to show both that the more heterogeneous organisms have been evolved in the later geological periods, and that Life in general has been more heterogeneously manifested as time has advanced.[30]

From then on 'persistence' made increasing inroads. For example, Spencer attended the Geological Society on 15 December 1858 to hear Huxley discuss the crocodile-like fossil *Stagonolepis*, one of the important reptiles dispatched from Elgin (Morayshire) by the Rev. George Gordon. Huxley's was an important paper, crucial in the long-standing debate over the age of the famous Elgin sandstones, whether Devonian or Triassic (Huxley opted for the latter, on account of this crocodilian fossil). Murchison, who needed convincing, was present, and so were Hooker and Spencer. As the meeting progressed Huxley not only likened the armoured *Stagonolepis* to a crocodile, but left the impression that crocodiles were therefore reptilian Methuselahs, an age-old group surviving virtually unchanged. The next day Spencer dropped Hooker a note, remarking that the "evening was a triumph for Huxley, and rather damaging for the progressive theory, *as commonly held*".[31]

This and Huxley's lecture on "Persistent Types" in 1859 had their effect. When Spencer reprinted the *Westminster* essay as a chapter in *First Principles* (1862) – which Huxley proof-read – he diplomatically rewrote the relevant section giving far greater emphasis to 'persistence' (or rather to the fact that progression could not be proved from fossils). Obviously worried, he insisted that "the opponents of the development hypothesis" could take little comfort from Huxley's words because an immense "pre-geologic era" had undoubtedly seen "biologic changes" going on "at their usual rates". And he still left a few lines at the end appealing to Owen's *Palaeotherium* argument, before hastily passing on to human history where progress *was* discernible. But piecemeal erosion continued with Huxley's anniversary address (1862), and in Spencer's *Principles of Biology* (1864) – again proof-corrected by Huxley – 'persistence' now swamped the discussion. Owen no longer warranted even a mention. Spencer agreed that "geologic" progress "may be, and probably is, illusory", that many fossil groups show "little or no progression", and that "the total amount of change is not relatively great, and that it is not manifestly towards a higher organization."[32]

Huxley did not entirely undermine the *First Principles*, for Spencer could still look back to a steady "pre-geologic" ascent. Yet there was an obvious incongruity. By Spencer's canons life *must* have progressed; increasing "heterogeneity" was the inevitable result of a multiplicity of effects flowing from every cause (the philosophic mechanism ensuring spiralling complexity). Not only was progress philosophically inevitable; evolutionary contingencies themselves made it a "liability". Discussing the environmental cause of life's ascent in *Principles of Biology*, Spencer argued that "*continual* changes in external conditions" call for new adaptations which are therefore "continually superimposed". Thus he concluded:

> While we are not called on to suppose that there exists in organisms any primordial impulse which makes them continually unfold into more heterogeneous forms; we see that a liability to be unfolded arises from the actions and reactions between organisms and their fluctuating environment.

Spencer's tragedy was of course that he could never kill a beautiful theory with an ugly fact. 'Persistence' was acknowledged, swallowed up, and ultimately obliterated by an overpowering progress. And this was characteristic of the American reaction; as New Englanders rushed to apply the *Synthetic Philosophy* in the Gilded Age, Huxley's idea of fossil persistence was swamped and Owenian progression taken to be truly cosmic in scope.[33]

Biogeography and Living Fossils

Logically, Huxley's position was compatible with Darwin's – progress was not inevitable by the canons of natural selection, and Darwin could easily accommodate living fossils like the marine bivalve *Lingula*, which has barely changed since early Palaeozoic times. But he too was worried by Huxley's weighting of the scales. Indeed it is possible that Huxley's geological doctrine led to areas of more widespread disagreement. 'Persistence' obviously reinforced the notion of relics – isolated faunae, often island cast-aways, cut off from mainline evolution. These archaic survivors trapped in the backwaters had given rise to the popular notion of living fossils. But as Huxley began to look at biogeography from the standpoint of his "persistent types", he took a radically different position to Darwin's. As we will see, he broke some of the geographical canons established in the *Origin*, highlighting once again the distinctiveness of his world view.

Presumably Huxley's need to prove 'persistence' sent him delving into the paradoxes of fossil and recent distribution. These issues he dealt with in his 1869 paper to the Geological Society "On *Hyperodapedon*" (another Elgin reptile, similar to the archaic living tuatara of New Zealand), and in his address as president in 1870, where one group brought the problem to a head – the marsupials.

In mid-Victorian times almost all Triassic and Jurassic mammals were thought to have been marsupial. The majority were likened to present-day Australian mammals, and the entomologist Andrew Murray was not the first to single out the marsupial ant-eater *Myrmecobius* as the nearest living relation of the fossil mammals of Mesozoic Europe. In late Victorian times this strange beast was known to be declining, unable to hold its own against the colonists' dogs, and it was already restricted to the south-west corner of the continent. Besides possessing more teeth than other marsupials, living or extinct, *Myrmecobius* is atypical in being pouchless (like American opossums). It was taken to be an "unmodified survivor" from Mesozoic times.[34] The question, however, was not how it had managed to survive; what really mattered was how it had evolved from antipodean parents in the first place.

Plant distribution posed a similar problem. The Viennese botanist Franz Unger had catalogued the resemblances between Europe's Eocene flora and the "prehistoric" gum trees of Australia, and being inclined to rule out the possibility of "several centres of creation", he assumed that *"Europe stood in some kind of connection with that distant continent"*.[35] Supposedly a land bridge between Europe and the antipodes in Mesozoic and Eocene times had allowed free passage. For the rest of the Tertiary European forms had progressed, while in the Australian backwaters these "prehistoric" animals and plants persisted as living fossils.

Gareth Nelson sees an "obvious parallel" between the mid-Victorian exodus from industrial Europe and contemporary faith in the "northern origins of the more dominant forms of life". Often it is impossible to miss this transposed colonialist image. The period was one of unprecedented expansion, with the dispossessed workers who had flocked to the factory towns being forced to emigrate to the colonies in search of a new life: the Welsh to Patagonia, the great exodus to Australia (where the population trebled in the 1850s, with four states being granted responsible government by mid-decade); the annexations in imperial India by the Marquess of Dalhousie from 1848–1856; and the influx into South Africa after the opening of the Transvaal goldfields and Kimberly diamond mines between 1869 and 1871. Europe was the active embarkation point. In the case of Australia, the ex-penal colony in the antipodes simply soaked up the northern hordes, just as in former times it had welcomed marsupial 'colonists'. After the Triassic period the continent was cut off and according to prevailing views the fauna simply stagnated.

Palaeontologists in Australia were naturally unhappy with some of these assumptions, noting that the Eurocentric views had palaeontological consequences which were at least partially testable. Frederick McCoy (1823–1899) had been appointed by John Herschel and Astronomer Royal G. B. Airy to the post of Professor of Natural Sciences in the new Melbourne University in 1854. He responded in 1862, trying to give the lie to "speculations based on the supposition that Australia, unlike the rest of the world, had remained as dry land since the Oolitic [Jurassic] period, and that the living little *Myrmecobius* and *Parameles* or *Bandicoots* were the associates of those little marsupials which lived" in Mesozoic times. Faunal 'persistence' was impossible because "the greater part of the country sank under the sea during the Tertiary period". He insisted with almost Agassizean severity that "every trace of the previous creations of plants and animals was destroyed and replaced by a totally different new set."[36] The Dublin-born McCoy was probably a Sedgwickian progressionist; he argued for a step-wise ascent of fossil life, and found this to hold as good for Australian as Irish rocks, although he utterly repudiated the notion of "development" by transmutation. He had actually worked for Sedgwick, cataloguing his Woodwardian fossils at Cambridge at the time (1846–1850) when the great geologist was besieging the hapless *Vestiges*. In the colony, McCoy located the Cretaceous beds and worked out the sequence of Tertiary strata, which paralleled those of the northern hemisphere. So far as he could tell, the entombed fauna had also undergone successive changes, although remaining peculiar to its type. Hence just as in South America the living sloths were preceded by the gigantic *Megatherium*, and in Ireland the deer by enormous elks, so in Australia the diminutive wombats were preceded by rhinoceros-sized diprotodonts and the small kiwi by huge moas.

Still civilised Europe, for its part, was quite content to view Australia as a faunal backwater, a kind of palaeontological penal colony. Huxley, for example, furnished evidence of a link between the exotic survivors of New Zealand and the Euro-American vertebrates of the Triassic. Numerous fossil footprints in the Connecticut Valley were attributed to wading 'birds' like the gigantic *Brontozoum* (with its six-foot stride), and likewise New Zealand was recently inhabited by the moas, "some of which were competent to keep stride with the *Brontozoum* itself." And at Elgin a reptile close to Huxley's heart, the *Hyperodapedon*, seemed in all respects "so extraordinarily similar" to the tuatara, a burrow-living, lizard-like reptile confined to New Zealand. Ignoring McCoy, Huxley asked in 1869:

> What if this present New Zealand fauna, so remarkable and so isolated from all other faunae, should be a remnant, as it were, of the life of the Poikilitic period which has lingered on isolated, and therefore undisturbed, down to the present day?[37]

To explain this migration and isolation, Huxley reworked some fashionable ideas. Since today's "distributional provinces" were inapplicable to the ancient past (something Darwin and, by this time, Wallace vehemently denied), he told the Geological Society in 1870 that "a vast alteration of the physical geography of the globe" must have occurred: a shifting relationship of land and sea such that Australia had long ago engulfed the European zone, allowing free migration. To expedite matters, he summoned up a lost "Mesozoic continent" somewhere in the north Pacific, making it the original breeding ground of the marsupials. From here they had migrated across the globe, continental extension by sea-floor upheaval providing convenient bridges to Europe, Australia and South America. The "Mesozoic continent" eventually sank without trace, and the marsupials died out in all their new homelands bar a couple, for the scenario could explain why Australia was still tenanted by what is "essentially a remnant of the fauna of the Triassic."[38]

None of the Geological fellows addressed by Huxley would have missed his drift. Lost continents abounded at the time (much to Darwin's disgust). Huxley's predecessor at the School of Mines, Edward Forbes, had called up a drowned Atlantis; and more recently, in 1862, Unger himself had invoked a mid-oceanic "Island" stretching as far north as Iceland, to transport American Tertiary plants to Europe. (Although it proved none too popular – most geographers preferred a "north-west passage", perhaps via Greenland or over the Pole, where the "climate at this time was genial".) Hooker, whose own voyage to the Antarctic on HMS *Erebus* under the Polar explorer Captain James Ross had provided first-hand experience of the flora of Tasmania, New Zealand and South America, told Darwin as early as 1851 that he was "becoming slowly more convinced of the probability of the Southern Flora being a fragmentary one – all that remains of a great Southern continent."[39] And Andrew Murray, in his handsome quarto *The Geographical Distribution of Mammals* (1866), had likewise employed a southern supercontinent, embracing Australia, South Africa and South America, to get over distributional problems.

Reactions to Murray were polarised and provide something of a litmus test. Huxley was duly impressed, calling Murray's volume "thoughtful and ingenious"; indeed he only differed in the more northerly siting of his own "Mesozoic continent". Darwin on the other hand never appreciated the man. In the first place Murray had been rather quick in attacking the *Origin* in 1860, giving Darwin a rather dim view of his abilities. But Murray now candidly admitted his haste and in his *Geographical Distribution* warmed to some of Darwin's views. He was still appalled by the nightmare of a world proceeding by "an infinity of experiments", mostly abortive, until the right one was hit upon accidentally. Rather, when the time was ripe for, say, a tree kangaroo, one "appeared perfect at the first attempt". However on points of geography he now generously admitted to be standing on Darwin's shoulders. But Darwin was equally horrified at his talk of sunken lands, remaining a staunch believer in the permanence of continents and oceans. In the *Origin* he had harangued those who demanded "prodigious geographical revolutions". He abhorred the mythical Atlantis and considered Murray's science a travesty, concluding from *Geographical Distribution* that "the man cannot reason", although agreeing with Hooker that "here and there clever thoughts occur." His lasting impression, he told Wallace as late as 1876, was that Murray had an "utter want of all scientific judgment." At the same time he protested to Wallace about those who sink "imaginary continents in a quite reckless manner". Rarely would he tolerate continental extension (the spreading of continental margins to embrace oceanic islands and allow their colonisation), although on this score even Hooker sided with the opposition. But then, as he told Darwin, "pretty nearly all the thinkers" are "on the continental side". Not that the camps were divided by hard and fast lines. There was a certain swaying to and fro; Hooker in the fifties had been fairly adamant on the need for a continent connecting New Zealand and South America, and Wallace was also an "extensionist". In the sixties Hooker wavered under a barrage of counter-arguments from Darwin (though never totally succumbing). But Wallace did defect. By 1863 he had joined Darwin, accepting the fixity of continents and dismissing Hooker's southern connection; and in the *Geographical Distribution of Animals* (1876) he pleased Darwin enormously by his poke at "those who would create a continent to account for the migrations of a beetle."[40]

By this time, though, Huxley had thrown in his lot with the "continentalists", lining up against Darwin and Wallace. Nonetheless there were still wide areas of agreement. Both Huxley and Wallace subscribed to what Murray considered the "generally adopted view", i.e. that in Mesozoic times the worldwide flora and fauna was "homogeneous", perhaps because of uniform global temperatures. In that case, marsupials – far from being localised – probably flourished on all continents, dying out everywhere except Australia and South America as a result of climatic change or competition. This belief had important repercussions. O. C. Marsh at Yale, brought up to accept worldwide distribution (he even nodded his approval of Huxley's "Mesozoic continent"[41]), not unnaturally expected to locate Jurassic mammals in the American West. He searched and identified his first specimen in 1878.

Huxley's drowned continent was also designed to serve the same end as Darwin's accidental means of dispersal (drift, wind, rafts, etc.). That is, to ensure an unbroken line of descent and thus effectively answer Agassiz, McCoy and the anti-evolutionists. It allowed a genetic link between isolated but closely allied species. *And* – something I dare say uppermost in Huxley's mind – it served to douse the flames of revolt among idealists. Huxley's student St George Mivart and another friend, Martin Duncan, not to mention the influential German idealist von Kölliker, opted for a rival polyphyletic approach: they believed that under some progressive law species from radically distinct stocks and even different parts of the world could and did converge in appearance. The connection between idealism and polyphyly will be explored in Chapter 6; here we need only note that idealists were bent on proving that similarities did not necessarily imply a common evolutionary origin. Hence Duncan suggested that Australian marsupials might have had a genesis quite independent of European Triassic ones.[42] He was *un*relating them, giving them separate lineages, and thus undermining the value of the Darwinian 'tree'. Huxley's and Darwin's dispersal mechanisms were designed to nip such heresies in the bud.

It is by no means certain that Huxley's continentalism grew out of his earlier theory of "persistent types". No such tendency manifested in Lyell, who sided with Darwin on continental stability and presumably found Huxley's sunken island equally unacceptable. And yet we can speculate on a direct link between lost lands and Silurian mammals. After all, a submerged continent would have provided a perfect hiding place for those desperately needed Paleozoic mammals and birds, no trace of which had ever turned up. Suddenly their absence would have seemed less embarrassing; they could have gone to a watery grave with their lost world.

Lankester's Social Strategy

Huxley's 'persistence' failed to capture the scientific imagination for a variety of social and theological reasons. Christians could have been expected to reject a steady-state geology and favour unidirectional change because it harmonised better with the historical progression from Creation → Incarnation → Judgement, and because a fossil ascent provided reassuring evidence of God's intent to introduce Man last on the earth. But even those who found God "Unknowable" or unthinkable in the 1860s adopted a progressive social philosophy, and many of Spencer's followers, especially in America, back-projected their social metaphysics onto evolutionary theory.

This is not to deny that 'persistence' could serve as a social *warning*, an aspect we might do well to concentrate on here. That "stormy petrel" Ray Lankester (1847–1929), who as a boy had hero-worshipped Huxley, consistently slated contemporary faith in universal progress, calling it an "unreasoning optimism". But his bringing 'persistence' back into the limelight was almost unintentional; at least, it was not an exercise in 'pure' science, more a by-product of his campaign to win professional recognition for scientists.

Ray was the son of a famous father, the microscopist Edwin Lankester, who had studied at University College (where Ray was to become Professor of Zoology in 1875). Despite being brought up in an Anglican household, he saw the comings and goings of such notables as Darwin, Hooker, Tyndall, and his "hero" Huxley; and although his father had strong reservations about the *Origin* – and Ray himself apparently suffered pangs on relinquishing his faith for the Darwinian creed – he nonetheless became one of the staunchest anti-teleologists of the day. His personal and professional ties with Huxley grew throughout life – as a teacher in London he introduced Huxley's laboratory techniques and type system, and, following Huxley's example, began his fossil studies with the armoured fishes *Cephalaspis* and *Pteraspis* from Devonian rocks. And when Huxley died Lankester poured out his feelings: "There has been no man or woman whom I have met on my journey through life, whom I have loved and regarded as I have him . . ."[43]

With Huxley his "father-in-science", Lankester was almost a second-generation Darwinian, and it is not surprising that he made good use of Huxley's doctrines. But by far the most intriguing was his appropriation of 'persistence' for 'political' purposes. In an increasingly technological society, worrying seriously for the first time about German competition, it served as an object lesson. Stagnation and degeneration were the fates awaiting any organism or society which failed to compete successfully – and by the later sixties the perils of lagging in the industrial race were becoming obvious. The scientists were the first to raise the alarm, and Lankester's little book *Degeneration* (the outcome of an address to the British Association in 1879) was a cautionary tale: Victorian doom-mongering at its plainest, designed to persuade Her Majesty's Government to mend its ways and endow science properly for the nation's sake. *Degeneration* starts off innocuously enough, as a study of parasites, with Lankester cataloguing the loss of limbs, eyes, etc., in animals cocooned within a host and glutted with food. He pointed out that not only might natural selection leave organisms "exactly fitted to their conditions, maintained as it were in a state of *balance*", it might actually operate "to diminish the complexity of [an animal's] structure."[44] Although this applied principally to parasites, and perhaps barnacles and ascidians (sea squirts), the moral was clear enough. Just as a "highly-gifted crab" might degenerate into a "mere sac" under conditions of plenty, so "an active healthy man sometimes degenerates when he becomes suddenly possessed of a fortune". And we have only to look to Rome to see an entire civilisation collapse "possessed of the riches of the ancient world." The warning in *Degeneration* was casually apocalyptic. But to see Lankester's ploy as part of a wider strategy, we need to know the social aspirations of the group he was championing.

Like his idol Huxley, Lankester was a publicist *par excellence*, exploiting the British Association and other platforms to campaign for recognition and state funding for science. (In modern parlance, he was fronting a pressure group.) The interests he was promoting were those of the emergent professional scientists; he was attempting to whip up interest in their cause on the grounds that the nation's strength depended on technical expertise. The Huxley/Tyndall/Lankester manifesto would have looked disquieting to an earlier generation. No longer was science portrayed as the handmaiden of natural theology. On the contrary, as nature became increasingly bloodthirsty, the wisdom and goodness of God were having to be sought in more tranquil moral realms. Nor was science linked so enthusiastically to the once-powerful self-help tradition. Before mid-century, *laissez faire* and entrepreneurial practice in industry were reflected by an entrepreneurial approach to science, and to indulge one generally had to be wealthy. The reformers now wanted it out of private hands and placed in the public sector. They wanted, in short, a nationally-organised professional community, equipped with modern research facilities, paid for and supported by the Exchequer.

The bubble had actually burst in 1867 after Britain's disastrous showing in the Paris exhibition. This created a "near-panic" situation as delegates reported home that Britain was losing her technological lead.[45] But even before this time, lobbyists like Huxley had seen the advantage of linking science to Britain's greatness. For example, in defending the *Origin*, he damned the pious "meddlers" who would stifle uncomfortable research, and predicted that the nineteenth century would "see revolutions of thought and practice as great as those which the sixteenth witnessed". England was destined to play a noble role in this "new reformation".

> She may prove to the world that for one people, at any rate, despotism and demagogy are not the necessary alternatives of government; that freedom and order are not incompatible; that reverence is the handmaid of knowledge; that free discussion is the life of truth, and of true unity in a nation.
>
> Will England play this part? That depends upon how you, the public, deal with science. Cherish her, venerate her, follow her methods faithfully and implicitly in their application to all branches of human thought; and the future of this people will be greater than the past.
>
> Listen to those who would silence and crush her, and I fear our children will see the glory of England vanishing like Arthur in the mist . . .[46]

There spoke a man who had known a decade of financial worry and hardship; who had struggled against odds to secure a job teaching science, only to find it as poorly paid as it was lacking in prestige. Linking science to social and technological salvation, almost in defiance of the other-worldly kind, was sound policy. The implication was that power and greatness lay in Truth, which meant in effect that the nation's health was materially linked to the well-being of the scientist. The argument never appeared quite so self-serving, and Lankester even denied "any suggestion of self-interest" after lodging his claim for State aid. Nor am I for a moment suggesting that it was a conscious motive, which would be to take a mean and cynical view. And yet the timing of events tells its own story; we should never lose sight of the fact that the movement first gained momentum in the fifties, when the scientists' financial plight must have seemed the greater for knowing that the economy was on the upturn and more money could be made available.

The product 'science' still had to be marketed. This meant generating a need for the professional services of scientists, which called for reorganisation. Their power-base had to be concentrated by breaking outside domination (or what was seen as domination), e.g. by driving a wedge between science and theology.[47] By circumscribing power, making science self-policing and raising standards, these professionalisers were forging a community along recognisably modern lines. Lankester's language deliberately portrayed research in a radically updated way, immediately understandable to *Times* readers in a labour-intensive age. Knowledge had become a commodity (this is what he was selling, for example, in his "Biology and the State" address to the British Association). He talked of *"creating new knowledge"* as if it were wealth or consumer goods,[48] and referred to the "occupations of making and distributing knowledge" as though research institutions were (or should be) intellectual factories: an industrial metaphor incomprehensible a generation or two earlier and which marks the emergence of modern corporate science. With 'knowledge' indispensable and funding assured, the scientific civil service would become both self-perpetuating and politically powerful.

Lankester got down to brass tacks in "Biology and the State", refusing to apologise for raising "such sordid topics" as the financial plight of British science, but intent on embarrassing the Government into action. He called it a national disgrace that "our English students flock to Germany" because of its superior research establishments and "well-trained army" of technicians and teachers. He pointed out that British aid to science was only a quarter of that enjoyed in Germany, even though it was a much poorer country.[49] His other tack was to employ contemporary ideas about evolution to highlight the unenviable fate should Britain continue on this disastrous course. *Degeneration* was both allegory and warning. On one level Lankester was merely discussing parasites; on another, he was actually articulating the aims of the young guard. He cautioned that just as innumerable past civilisations had collapsed, so the threat of stagnation or decay hung over "the white races of Europe" today. As befits Huxley's heir, he called the "tacit assumption of universal progress" a fool's paradise, and suggested that "naturalists are so generally persuaded" of its universality "by habit, not by reason".[50] By urging that we "are as likely to degenerate as to progress", Lankester sought to establish a kind of paranoia, the social malaise for which he – or his group – had the remedy. This was the "full and earnest cultivation of Science", nothing less could save Britain from the fate of our "ruined cousins – the degenerate Ascidians".

Thus the concept of a stationary or back-sliding evolution was an integral part of the professionalisers' strategy. One might even speculate that because Huxley's doctrine found such good polemical use, by a curious dialectical process, its scientific status was thereby increased.[51] After all, the vociferous campaigners who exploited 'persistence' were the ones likely to benefit most from new government initiatives, which would in turn enhance their role as official spokesmen on palaeontological matters.

4

Social Function and Fossil Form
A Case Study: The Dinosaur
(1841–1875)

[The generation of knowledge] cannot be understood in terms of psychology, but must be accounted for by reference to the social and cultural context in which it arises. Its maintenance is not just a matter of how it relates to reality, but also of how it relates to the objectives and interests a society possesses . . .

Barry Barnes, *Interests and the Growth of Knowledge* (1977).[1]

New Perspectives

Maintenance or exploitation of knowledge is only half the problem. Concentrating on, say, Lankester's none-too-subtle use of 'persistence' might leave the impression that the doctrine itself was originally derived by 'impartial' means, by a disinterested scientist going about his business. The fact is, Huxley might have devised or modified the theory to meet his own needs in the first place. Indeed, there is circumstantial evidence for this, in particular we know that he was subject to covert pressures and social influences, of the sort that might have predisposed him to a theory of "persistent types".

In the present chapter I want to examine this aspect of the creative process, focusing on the wider social concerns which have helped to shape palaeontological thought, while trying not to lose sight of the *scientific* worth of the final product.

Barry Barnes talks about "ideas as tools with which social groups may seek to achieve their purpose . . ." Not only abstract ideas, but ones embodied in concrete form, as in the case of dinosaur reconstructions, make equally effective "tools". In many ways the dinosaur – the *Victorian* dinosaur – is a singularly appropriate choice for study. Its restoration often entailed free use of the imagination – although, obviously, imagination is anything but 'free'. It is socially and culturally constrained in a myriad of unrecognised and unspecified ways, and the creative component in any reconstruction may be the door through which ideological influences enter to shape the beast.

Numerous ways now exist to tackle the generation of knowledge in its proper historical context, but sociology of knowledge is, for our purposes, the most interesting. It acknowledges that broader social concerns – for instance the political, religious, class, or racial leanings of the group to which the scientist belongs – may actually become interwoven in the fabric of his theory, giving it colour and shape, if not meaning. It is certainly suggestive that supposedly objective statements often correspond as much to contemporary needs as 'reality'. Now, the operation of interests in science is a prickly point, and the hardest thing, as Rudwick wrote in 1980, will be to convince those whose very livelihood seems to depend on expunging every trace of prejudice – working scientists themselves. But Barnes has treated the problem with characteristic sensitivity. Indeed, he has complained that some "existing work leaves the feeling that reality has *nothing* to do with what is socially constructed." Rather, he suggests that:

> Knowledge is not related to activity rather than reality; it is related to activity which consists precisely in men attempting to manipulate, predict and control the real world in which they exist. Hence knowledge is found useful precisely because the world is as it is; and it is to that extent a function of what is real, and not the pure product of thought and imagination.[2]

Where before, cognitive sociology tended to languish on the periphery, explaining deviation from 'true' science in terms of biases, prejudices, and other distorting factors, a number of historians have now carried it into the citadel itself, attempting sociological explanations of the generation of so-called 'true' theories. This approach has never lacked critics, but judging by Steven Shapin's impressively-documented forthcoming paper in *History of Science* (1982), it is looking increasingly attractive to scholars from a number of disciplines who are no longer content with the rational reconstruction of ideas divorced from context.[3]

In this relaxed atmosphere, an attempt to relate the various guises of the Victorian dinosaur to its changing 'role' might be welcomed. This avenue is certainly worth exploring, especially as the twists and turns taken by the various restorations often seem little related to new fossil finds. This fact alone suggests that the climate in which the dinosaur was reared might better explain its resulting shape.

Owen, Grant, and the Materialist Threat

The dinosaur was not born in innocence, nor was it the inevitable consequence of new fossils (despite the impression left by a number of Whiggish studies). The megalosaur and iguanodon bones that eventually went to make 'The Dinosaur' had been known since the twenties, yet the concept *Dinosaur* only became tactically useful a decade or so later, and Owen did not coin the word until 1841. This early episode falls outside the time span of this book, although its bare details are essential for an understanding of Huxley's reaction in the 1860s. Therefore what follows is rather more a quick impression gleaned from largely unpublished sources than a rigorous demonstration.

If we look to the thirties, i.e. *before* Owen's archetypal theology made fossil continuity an imperative, what stands out from his lecture notes and papers on the ape, platypus, and dinosaur, was his preoccupation with refuting Lamarckism: the belief that nature is an escalator, carrying life endlessly upwards. By this doctrine, chimpanzees might become men, with the ensuing loss of both dignity and reason as mankind submerged in 'irrational' nature. Owen was not alone in his denunciation; like Lyell and most of his generation, he desperately feared for the status of man in this debased scheme. Nor was it solely Oxbridge-educated Tories and country clergymen who saw French godless materialism at the root of the Revolutionary Terror, and atheistical philosophy the cause of social discontent. With Parisian insurgents again at the barricades in 1830, and agitation by radicals and republicans continuing until 1836 (the results of which Owen witnessed on his first trip to Paris in 1831) an understandable shudder crept through establishment science at home, leading to stabilising measures – an integral part of which was the identification and isolation of 'atheistical' Lamarckism.

Owen truly believed Lamarckism ill-founded, dangerous, and ultimately incomprehensible in its ungodly aspect. Like most of his generation he loathed its apparent materialism, which he equated with lack of providential control. And it was just this heretical aspect that Parisian-inspired anatomists and reformers, particularly Robert Grant in London, were championing. Robert E. Grant (1793–1874) was the epitome of everything Owen hated. To his contemporaries he must have appeared an out-and-out materialist; he believed in a lawful nature, but not one under Providential control. He was quite scathing on the subject, actually telling jokes about Providence to the acute embarrassment of his students.[5] Admittedly this was after 1860, but it gives some idea of his persuasion. His impact might have been lessened, or the man himself ignored, had he not been Owen's social rival to boot. Grant, a Scot who had graduated from Edinburgh, had taken the Chair of Zoology at the newly-founded London University in 1828 (a post he held until his death in 1874). He is an intriguing and shrouded figure, and every effort to resuscitate him is hampered by the disappearance of his bound volumes of letters. We know that at Edinburgh in 1826 he interested the teenage Darwin in microscopic life, and taught him the techniques later to prove useful on the *Beagle* voyage. Darwin found him stuffy "but with much enthusiasm beneath this outer crust". True, events show him to have had a fiery and uncompromising nature, despite appearances. Later, students in London were "strongly disposed to laugh at his peculiarities", for example his habit of turning up formally attired in swallow tails and choker, and they complained that his lectures were equally obsolete.[6] But in 1830 he had given possibly the most advanced course in comparative anatomy outside France or Germany. Like Owen, Grant played an active role in the infant

Zoological Society, delivering the first lecture series on comparative anatomy to its members in 1832–3. But about this time things turned sour, and one can only speculate that his ungodly philosophy, either tolerated or unvoiced in Edinburgh, made him unwelcome. He seems to have been ousted, first from the Zoological Society Council (Owen, to his "eternal disgrace", vetoing Grant's appointment[7]), finally vanishing from the Society's register altogether.

Grant's threat was not wholly science-based. He flouted convention and was openly provocative, practically relishing the air of unrespectability which surrounded him. He never married and is thought to have been a homosexual[8] (a problem compounding any interpretation of his social predicament). No "social experimenter" could reasonably expect to retain him as a friend and move in more elevated circles, and in the thirties Owen gave every indication of

Robert Grant aged 59, from a lithograph dated 1852. (National Library of Medicine.)

refusing to be beaten into second place by a rival of dubious character and obnoxious views. A rival Grant truly was. Owen was frequently exalted as the British Cuvier, yet the title was first bestowed on Grant, Owen's senior by eleven years. In the thirties, Grant had been tipped as the man most likely to succeed, particularly by *The Lancet* – a journal then agitating for medical reform, and firing shot at the Government, College of Surgeons and quackery alike. If one looks for Grant's power base, this was largely it. He mixed with Radicals, notably *The Lancet*'s hell-fire editor Thomas Wakley (1795–1862), a surgeon who spent a large part of his life fighting libel suits. Wakley publicly backed Grant, publishing his lectures and reporting his movements – and of course it was Wakley who condemned Owen for standing in Grant's way at the Zoological Society. He also tipped Grant as the new Cuvier.[9] It wasn't exactly a happy title. While Grant had been one of the Baron's familiars, vacationing in Paris at Cuvier's invitation, he was more in agreement with that *enfant terrible* Lamarck. Grant may even have attended Lamarck's lectures on invertebrates. And certainly he knew the other thorn in Cuvier's side, Geoffroy St Hilaire. Not surprisingly, then, Grant was instrumental in importing Lamarck's and Geoffroy's ideas on transmutation into Britain during the 1820s, i.e. *before* Lyell made Lamarckism a fighting issue by his polemic in *Principles of Geology* (1832).

A picture is emerging. Grant was uncompromising, courageously so, and largely supported by reformers and radicals. In a sense, his real place was Paris, where he might have been accepted with less fuss. However even in London, during the early years, his social position had strengthened – at the University, Zoological Society and Royal Society – exactly paralleling Owen's rise. At first they had been intimates, but relations were to cool as Owen took steps to neutralise Grant's threat. A single act, like vetoing his Zoological Society post, effectively stripped Grant's power in that institution – and greatly increased Owen's opportunities for original research on exotic cadavers. (The results of these dissections and the ensuing string of papers actually made Owen's reputation.) Grant's materialism posed a moral and social threat, and Owen's counter-measures were an integral part of his larger responsibility as a Broad Church scientist. He guided his research along lines which reinforced the status quo: making ape anatomy inaccessible to Lamarckians, dismissing alleged embryological evidence for transmutation, and so on. Even his work on the platypus, which gained him a coveted FRS, was an attempt in part to counteract Grant and Geoffroy – to show that the strange duck-billed mammal (which Grant believed laid eggs) was not transitional at all, or bird-like in the way demanded by Geoffroy and his supporters. The dinosaur was another part of this grand strategem – a natural extension of Owen's ideological programme into the fossil arena.

By designing the dinosaur Owen was doing what was expected of him. In palaeontology there were recognised ways of dealing the death blow to Lamarck – and as a young anatomist making his way up the social ladder Owen could certainly gain powerful allies by endorsing these procedures. Geologists like the Oxford catastrophist William Buckland were already using retrogressive fossil sequences; Owen was now to follow suit and deploy fossil reptiles in such a way as to prove that a necessary ascent according to some Lamarckian law was nonsense. He took Buckland's "Great Lizard" (*Megalosaurus*, unearthed in Oxford slates) and Mantell's *Iguanodon* and *Hylaeosaurus*, and created a wholly new order, evocatively called the "Dinosauria". The order gave them a new-found status, and separated megalosaurs from lesser reptiles (they had previously been seen as glorified lizards). Launching the revamped model at the Plymouth meeting of the British Association in 1841, Owen revealed sweeping design changes. Rather than sprawl, the "dinosaur" was to stand like a rhinoceros: it was a compact, multi-ton "pachyderm" – in fact, we know exactly how Owen envisioned them because thirty-ton concrete and steel models were built to his specifications in the Sydenham grounds of the Great Exhibition in 1854. Not only did Owen's dinosaur stand erect, but he placed it next to mammals on life's scale: i.e. he acknowledged dinosaurs as the highest reptiles, and equipped them with mammal-like four-chambered hearts and near-perfect circulatory systems. In other words, he was cunningly using Lamarck's cardiovascular criteria to place dinosaurs on the top rung. Having done this, he was able to turn Lamarck and Geoffroy on their heads. There was no unabated ascent, as Grant had insisted in his lectures (and Owen singled them out as betraying a heretic in their midst). Rhinocerine reptiles, rulers of the Mesozoic planet, had on the contrary degenerated into a sorry "swarm" of lizards.[10] At this early stage, Owen pictured a step-wise progression of life, punctuated by retrogressive fits, something totally inexplicable in terms of Lamarck's blind upward-striving forces.

The concept of "degradation" appealed to scientists across a broad theological spectrum, and enjoyed considerable vogue in the forties. The Scots geologist and Free Church leader Hugh Miller used it effectively against the *Vestiges*, so offensive to him because it

The *Megalosaurus* reconstructed with a rhinocerine aspect at Crystal Palace. (Photo: the author.)

denied the miraculous. While at Cambridge Sedgwick employed it because transmutation seemed to demolish Personal design in nature. The fact that the dinosaur was so profitable would have reinforced Owen's conviction of its soundness: it supported a higher metaphysical truth, thus it was a piece of *good* science. And good science is rewarded. By the forties, Owen was socially ensconced and the most influential comparative anatomist in the country – much of which was to do with his incredible industry harnessed to socially-acceptable ends. This is not to deny that inestimable other factors contributed to his success; equally, it would be hazardous to explain Grant's financial crash as simply the result of his self-destructive radicalism. Grant was in dire straits by this time, penniless and continually pestering the College authorities for a loan. At one point he was reduced to living in a slum in Camden Town amid "harlots and knaves", and running the risk of eviction during the police raids.[11] Numerous other factors contributed to his plight, not least University College's inability to make his lectures compulsory for medical students. Hence very few attended, and since his finances depended solely on student fees, he ended up earning barely £100 in good years, and less than £50 in bad ones.

With poverty and poor living conditions came a rapid decline in his scientific output, causing Darwin to note in his *Autobiography*, "he did nothing more in science – a fact which has always been inexplicable to me". He turned into a curiosity. Forbes, gossiping with Huxley in 1852, referred to him as "that shadow of a reputation" and asked what new "eccentricity" had manifested.[12] But the unkindest cut of all came from Huxley himself. One might have imagined the post-Darwinian generation, rewriting history their way, would have treated Grant kindly. After all, Huxley was first to admit that Grant was the only man he knew who had actually taught "evolution" before Darwin. But even the new pantheon was to elude him. Huxley – a very proper gentleman in the Victorian sense – probably saw Grant as much a social reprobate as Owen had done, and sourly commented that "his advocacy [of evolution] was not calculated to advance the cause."

Huxley and the Bird Paradigm

> . . . I am engaged [in] a revision of the Dinosauria, with an eye to the "Descendenz Theorie". The road from Reptiles to Birds is by way of Dinosauria to the Ratitae [flightless birds].
> Huxley to the German monist
> Ernst Haeckel, 21 January 1868.[13]

Owen's rhinocerine model – as an 'establishment' or anti-transmutatory bulwark – served perfectly well for a quarter of a century. Yet this period witnessed profound changes, and Owen himself underwent an Archetypal conversion, which affected his whole science and made fossil continuity a new imperative. Towards the end of the period the comparative quiet of English science, compared to, say, the frenzy in Germany during the fifties (whipped up by Carl Vogt, Ludwig Büchner and kindred church-baiting materialists), was finally shattered by Darwin's *Origin* and the higher Biblical criticism of *Essays and Reviews* (1860). Now England too was to witness a materialist upsurge which refuelled the campaign against Platonism and its "mystifications".

By the 1860s the social basis of science was in a process of being transformed, and it will pay us to look closer at the changing ethos if we wish to understand the palaeontological gambits of Huxley and his opponents. Modern historians like Roy MacLeod and Frank Turner, who have made a study of the new professionals, point out that most of the young guard had "grown up on the peripheries of the English intellectual establishment",[14] having been educated in Edinburgh, the London colleges and medical schools, or in Germany. Thus they lacked all Oxbridge privileges and as Turner says "possessed no ready access to the higher echelons of Victorian society." (Only 'second-generation' professionalisers, like Lankester, returned to Oxford – and complained bitterly of its medievalism.) In the fifties the struggling scientists, including Huxley (educated at Charing Cross Hospital) and the outspoken Royal Institution physicist John Tyndall (educated at the University of Marburg), found themselves almost second-class citizens. Yet by the 1870s this group – through their polemical demands and solid middle-class support – had captured a number of major institutions in the city: university chairs, society posts and new journals (like *Nature*). This stunning social transformation brought sweeping changes in the ideological make-up of science. Tyndall's outrageous claims acted like a visionary pointer to a changing age. He teased his "working men" with bold talk of babies built from chemicals, and other free-wheeling fictions, such as the thinking capacity of robots. Even Huxley began investigating human automatonism and "the mechanical equivalent of consciousness".[15] The movement had all the hallmarks of an evangelical revival. Tyndall's *Fragments of Science* was harnessed by the Rationalist Press, which issued 30,000 copies at rock-bottom prices – outdoing even the pamphleteering Methodists in their drive to reach Everyman. Huxley's inflammatory lecture "On the Physical Basis of Life" in 1869 likewise caused a sensation. Sales of the *Fortnightly*, which carried it, beat all records and the issue ran to seven editions.

The quasi-religious dimension of the movement was quite evident. "The ethical trenchancy of the Evangelicals was passing over to the agnostics," observed G. M. Young. Huxley, that "Roundhead who had lost his faith", in the words of the *Edinburgh Review*, crucified his theological enemies for the sin of faith and coined the word "agnostic" for himself, while insisting that Morality ultimately rested in acquiescence to the physical fact. All of which caused one commentator to presume that Huxley "left many of his readers wondering whether the agnostic scientist was not a more faithful heir of the heroes of the Reformation" than his Protestant contemporary.[16] Huxley deliberately steered clear of "the cynics who delight in degrading man" and "the common run of materialists, who think mind is any the lower for being a function of matter". Nonetheless he had the same missionary zeal, and carefully retained the Old Testament benefit of a "scientific hell, to which the finally impenitent, those who persist in rejecting the new physical gospel, might be condemned."[17] If anything, Tyndall was more daring; with an Orangeman's logic he had cut away the foundations of his own Protestantism, and continued hacking until he hit a bedrock of Matter. In his breathtaking cosmos, man and machine were subordinated to the same unyielding necessity. He crystallised its beauty and stark inescapability before the British Association in 1870: "at the present moment", he said, "all our poetry, all our science, all our art – Plato, Shakespeare, Newton, and Raphael – are potential in the fires of the sun."

These scarcely veiled mechanistic and reductionistic views, often urged with holier-than-thou and almost sectarian intolerance, were what really galled Romantic "liberals" like Owen and St George Mivart. It hardly helped to be told by Huxley that "The advancing tide of matter threatens to drown their souls . . ." Mivart, for one, was endlessly contemptuous of Tyndall's "new creed – 'I believe in One Force'." And though Owen had advised his YMCA listeners to relinquish their childhood myths and embrace a lawful creation, he would have been appalled by Tyndall's naturalism, extending as it did to man's very soul.[18] So the idealists had powerful ideological reasons for hating Huxley and his supporters, and later in this chapter I will show that their palaeontological ideas take on a new significance once we see them as a reaction to what were viewed as dangerous materialist trends.

But equally, we can better appreciate Huxley's actions by treating them from the standpoint of this 'sectarian' split. As part of his campaign to break up obstacles to a materialist evolution between 1867 and 1870, he began dismantling Owen's rhinocerine – and anti-transmutatory – dinosaur. In a series of lectures to the Royal Institution and Royal and Geological Societies, he subverted Owen's original intention, throwing the reptile into an evolutionary light and making it the ancestor of birds. This renewed attempt to bridge the classes (Owen, of course, having tried it with *Archegosaurus*) was timely and necessary: a number of geologists – Buckland's successor at Oxford, John Phillips, and even Lyell in 1859 – thought that evolution might only apply to species, leaving higher categories like families and orders untouched. On the contrary, Huxley hoped to establish that it could transmute one *class* into another. But first he had to overhaul the dinosaur completely; Owen's model was useless for his purpose, fit only to be disassembled and rebuilt to avian specifications. And this Huxley set about doing with appreciable relish.

How Crucial was *Archaeopteryx*?

The extent to which Huxley's actions were dictated by new evidence is debatable. Certainly some fascinating fossils had come to light, none more celebrated than *Archaeopteryx*. But this area is in need of thorough re-examination, since we have slipped into the habit of reconstructing Huxley's position as if he were a 'modern', expecting him to view *Archaeopteryx* as we ourselves might. Oddly enough he was relatively unimpressed by it. Not that the academic world was so generally unmoved. Henry Woodward in December 1862 called it "startling" and a "wonderful discovery", S. J. Mackie of *The Geologist* saw it as a "sensation", and John Evans used the epithet "marvellous". Nor did its possibly heretical usage escape unnoticed. At the Natural History Museum in Munich the ageing Andreas Wagner – in the first paper published on the feathered reptile – tried specifically to "ward off Darwinian misinterpretations". Wagner had only recently retired from a fierce debate with the staunch socialist and materialist Carl Vogt, and was critically aware of the danger should the feathered reptile fall into Darwinian hands. As a precautionary measure, he published a hearsay description in 1861 (never having seen the fossil himself), christening the beast *Gryphosaurus*, and insisting that it was an anomalous reptile totally unrelated to birds. Nonetheless, he anticipated that Darwinians, in their "fantastic dreams", would mistakenly welcome it as an "intermediate" creature. So the feathered reptile was a *cause célèbre* even before it was officially described. Huxley's intimates were also aware of its significance. John Evans, manager of the Nash pulp mills and an expert on ancient coins and flints (a mercantile combination which appealed to Huxley) wrote in Huxley's *Natural History Review* that in its "anomalies" *Archaeopteryx* tends "to link together the two great classes of Birds and Reptiles", and that therefore "Its extreme importance as bearing upon the great question of the Origin of Species must be evident to all".[19]

It is impossible to deny that "the geological and palaeontological worlds were astonished" at the discovery, as Mackie said. Wagner's polemic was translated and published in the *Annals and Magazine of Natural History* in April 1862, followed by von Meyer's description of an isolated feather from the same Solnhofen region. *Archaeopteryx*'s owner (a local doctor) hoped to push up its price by allowing visiting scientists to glimpse the relic without taking notes. Owen's negotiator, George Waterhouse, was dispatched to Pappenheim, Bavaria, and establishing the fossil's "true ornithic nature" bought it on behalf of the Trustees of the British Museum for £450 (an exorbitant sum which further added to its notoriety). It arrived in Bloomsbury in October 1862 and Owen wasted no time reading his description before the Royal Society. On November 20 he declared it "unequivocally" a bird, although of a distinct and primitive order. Its peculiarities were coming to light in a piecemeal fashion. Wagner had concentrated on the long bony tail, to which feathers were attached; Woodward, the young assistant in the Geology Department of the British Museum, also reported in the *Intellectual Observer* that the tail was "the great and striking peculiarity". Owen recognised this obvious "departure" (in fact, until the last moment he was set to call it *Griphornis longicaudatus* – "long tailed enigmatic bird"). But his reconstruction had shown up four unfused and clawed fingers – precisely what he might have predicted on his archetypal theory. *Archaeopteryx*, he announced, showed a "closer adhesion to the general vertebrate type", its "general" features, i.e. fingers and tail, being only "embryonal and transitory" in today's specialised birds. Many observers fell in with Owen's diagnosis: according to Mackie "Archaeopteryx presents an example of the

persistence of embryonic or more general characters". Although
Woodward noted that a number of "very distinguished naturalists"
were not so sure.[20]

More peculiarities showed up. Evans pointed to suspected cranial
bones "which appear to have escaped Professor Owen's notice",
and even to an impression of the brain cavity. He also thought he
could identify a beak with four teeth – although Owen dismissed
this as belonging to a fish. At the time, though, Evans was solidly
supported. Von Meyer, contacted for his views, concluded "that
the jaw really belongs to the *Archaeopteryx*". Mackie and Charles
Carter Blake teamed up on hearing of this and confirmed that "the
fossil brain" and portions of the skull did remain on the slab. Lyell
too thought it a "tolerable certainty", while Hugh Falconer, gleeful
perhaps at seeing Owen tripped up, waylaid Evans:

> Hail, Prince of Audacious Palaeontologists! Tell me all about
> it. I hear that you have to-day discovered the *teeth* and jaws of
> the *Archaeopteryx*. To-morrow I expect to hear of your
> having found the liver and lights! And who knows, but that
> in the long run, you may get hold of the *fossil song* of the same
> creature, impressed by harmonic variation on the matrix.[21]

Between Wagner's polemic in 1861 and Evans' controversial claim
for brains and teeth in 1865, *Archaeopteryx* was accorded an astonishing reception – which makes Huxley's ambiguous stand the
more perplexing. Wasn't Mackie referring to Huxley when he
observed:

> Those palaeontologists who were silently present [when
> Owen read his paper] at the Royal Society's meeting, or who
> were "conspicuous by their absence," whose opinions we
> should have been glad to know, have maintained a significant
> silence. And the practice of naturalists in this respect seems
> nowadays like the practice of superior officers in Govern-
> ment establishments, – to find fault whenever they can, but
> never to give any praise.

Prophetically true if the finger was pointing at Huxley, because
when he did tackle *Archaeopteryx* in 1868, it was mainly to emphasise Owen's "errors"[22] – in particular his inability to tell the right

foot from the left. Generally, however, Huxley's attitude towards the Jurassic bird is deceptive, and it is too simplistic to see him as a 'modern', as many accounts do; indeed, there could have been compelling reasons for his 'failure' to grasp the creature's latter-day significance, and it makes more sense to treat him from a contextual point of view.

True, Huxley was in no doubt about a reptile-bird relationship. Since 1863 he had been teaching students that birds were "merely an extremely modified and aberrant Reptilian type", and the chicken and lizard (or the stork and "snake it swallows") even became unlikely bedfellows in a new vertebrate "province", which he called the "Sauropsida". This *was* an unlikely combination (Owen had categorically denied it), so Huxley took a calculated risk. With exquisite timing, he casually proposed that dinosaurs – the founders of the avian dynasty – probably had "hot blood" and a bird-like heart and lungs.[23] (Sedgwick's young assistant at Cambridge, Harry Seeley, had suggested "hot blood" for pterodactyls, which might have given him the idea.) It was a speculative master stroke, even if

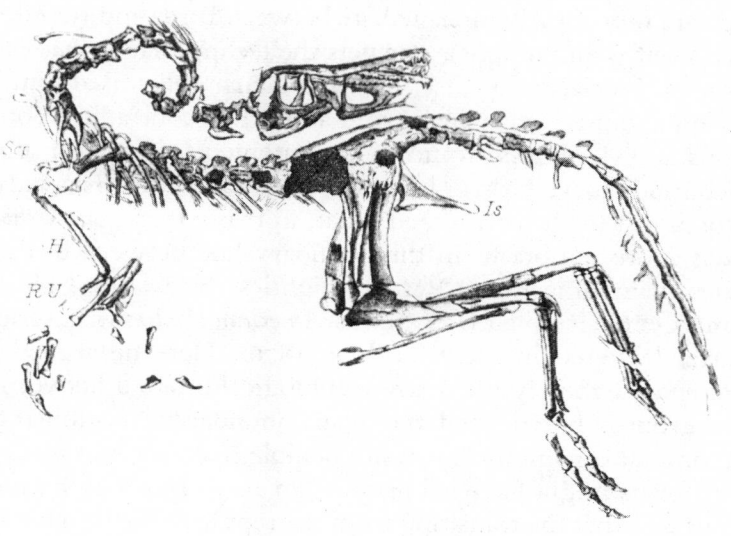

The small dinosaur *Compsognathus* with long, bird-like hind limbs. (From Huxley [1868a] in *Popular Science Review*, 7, 244.)

it did set off a whirlwind of controversy (see Chapter 6). Hot blood would have brought dinosaurs – and by association reptiles – closer to birds than any amount of turgid osteological comparison.

But this is not to say that he found *Archaeopteryx* necessarily transitional. He might have, had he been properly Darwinianised. In the 1860s, however, he was still committed to his neo-Lyellian 'half-way house'. He still accepted a "pre-geologic" date for all major evolutionary breakthroughs; for him, remember, birds were already flourishing in Palaeozoic times. So he could never have accepted Woodward's claim that the Jurassic *Archaeopteryx* "represents perhaps one of the very earliest examples of its class". It was far too young. Two alternatives were available to Huxley: either 1) *Archaeopteryx* was *not* an immediate descendant of reptilian stock, or 2) it was itself a "persistent type", coming down unchanged from the time of the real Palaeozoic transitional forms. On other occasions Huxley certainly resorted to this second type of option; we will shortly see that he used it to explain how Jurassic and Cretaceous *dinosaurs* could be ancestral to birds. But he apparently failed to apply it to the feathered fossil. This leaves the first option. Looking at all his papers (not just his famous "On the Animals which are most nearly intermediate between Birds and Reptiles" – which itself is ambiguous), one gets the feeling that *Archaeopteryx* was never crucial to his case, nor that it first galvanised him into thinking about an avian ancestry. I admit this contradicts popular notions and, I confess, some of his statements do appear almost self-contradictory. Others however are clearer: on first studying the fossil in 1868 he considered "that, in many respects, *Archaeopteryx* is more remote from the boundary-line between birds and reptiles than some living Ratitae [flightless birds] are." He only becomes explicit much later, in his *American Addresses* (a series of lectures delivered in New York in 1876). Here he argued that *Archaeopteryx* merely filled a morphological interval between existing groups. It stretched the avian boundaries, demonstrating "that animal organisation is more flexible than our knowledge of recent forms might have led us to believe . . . But it by no means follows . . . that the transition from the reptile to the bird has been affected by such a form as *Archaeopteryx*."[24] It was an extreme avian variant, an "*intercalary*" type he called it, and he distinguished this from the "*linear*" or truly transitional type.

The oldest known fossil bird was seemingly irrelevant to Huxley's case; for him "the *Ratitae* are the scanty modern heirs of the great multitude of ornithoid creatures which once connected Birds with Reptiles." But given Huxley's neo-Lyellian views this makes sense. His theory of 'persistence' predisposed him to look for evidence of the *oldest* birds, and however old or incontestible *Archaeopteryx*, the Connecticut Valley trackways were inestimably older (Triassic, in fact). By the 1840s, this famous footprint site had yielded forty-nine "species" to Edward Hitchcock, Professor of Natural Theology and Geology at Amherst College, thirty-two of which indicated bipedal "waders". Some were huge: *Brontozoum giganteum* had an eighteen-inch print and a six-foot stride. Hitchcock was still scouting and reshuffling his "species" well into the Darwinian era. By 1863 he had counted over 20,000 individual prints and while dropping "ten or a dozen of my old species of footmarks", he had described some 30 or so new ones.[25] Huxley obviously endorsed this evidence of late Triassic bird life. To be fair, he actually envisioned a mixed community, seeing both moa-like birds and dinosaurs wading the shores together in Triassic times (although by 1877 Marsh was convinced that dinosaurs alone had been responsible for the tracks). Triassic ostriches, then, fulfilled a neo-Lyellian need, but they also made *Archaeopteryx* a late-appearing anachronism, almost useless for Huxley's purpose – it ruined the neat line he had drawn connecting tall struthious birds with strutting dinosaurs.

Strutting dinosaurs – here at last was fossil evidence that Huxley did not ignore. The finds in New Jersey of bipedal dinosaurs, both herbivorous (*Hadrosaurus*) and carnivorous (*Laelaps*), were crucial to his case. Joseph Leidy, professor at Pennsylvania University and doyen of American vertebrate palaeontology, entertained little doubt that these American dinosaurs sprinted on their hind legs (especially the "leaping" *Laelaps*), and his former student Edward Drinker Cope had them reconstructed in bipedal poses for a projected Paleozoic Museum in Central Park in 1868 (a project scrapped when the models were vandalised by hoodlums, under orders from the infamous Tweed Ring in 1871). Huxley gladly incorporated this evidence in his celebrated Royal Institution lecture on avian origins in 1868. He assumed that *Iguanodon*, like its American cousins "walked, temporarily or permanently, upon its hind legs". Here, then, were "extinct Reptiles which approached these flightless birds [ratites], not merely in the weakness of their fore-limbs, but in other and more important characters." Even so, size was a major headache; few of the American and British dinosaurs fell short of thirty feet, which made them unlikely candidates for a bird ancestry. This partly explains why Huxley placed greater emphasis on another Solnhofen fossil, *Compsognathus*. Wagner had first described it in 1861, although failing to include it among the dinosaurs; Haeckel's colleague at Jena, Carl Gegenbaur, had noticed its bird-like anatomy, but it was really left to Cope and Huxley – who simultaneously alighted on this delicate reptile – to push it into the limelight. For Huxley it afforded "a still greater approximation to the 'missing link' between reptiles and birds." First by its accommodating size, being little more than two feet, but also because of its long delicate hind limbs, which made it impossible "to doubt that it hopped or walked, in an erect or semi-erect position, after the manner of a bird".[26] So Solnhofen *did* ultimately provide the 'missing link', or as near as one could get, but Huxley had groomed *Compsognathus* – not *Archaeopteryx* – for the role.

Timing was a critical problem. These dinosaurs were all Jurassic or Cretaceous; according to Huxley's chronology, the actual transition must have taken place innumerable ages beforehand. To retain some shred of consistency, he had to tack on an *ad hoc* clause. In his 1870 address to the Geological Society, he admitted it "very doubtful" whether these dinosaurs were "the actual linear types by which the transition from the lizard to the bird was affected." The real ancestors were probably "hidden from us in the older formations." He repeated this message during his American tour of '76, telling New Yorkers:

> It is, in fact, quite possible that all these more or less avi-form reptiles of the Mesozoic epoch are not terms in the series of progression from birds to reptiles at all, but simply the more or less modified descendants of Palaeozoic forms through which the transition was actually affected.[27]

Huxley's position overall was a tricky one. He was skilfully juggling potential fossil proofs of evolution, while trying to make them conform to his oddball timing of events. So we should beware of making him a 'modern'. Being deeply enmeshed in contemporary issues, he was subject not only to Darwinian and neo-Lyellian influences, but equally potent anti-Platonic and anti-Owenian ones. He was building an "avi-form" reptile both to plead Darwin's case and discredit Owen's 'mammalian' – and anti-evolutionary – rival. But construction was only half the battle, he still had to convince colleagues of its viability; and since a number were positively antagonistic to Darwin's "higgledy piggledy" selection, he approached them personally with assurances that an "avi-form" dinosaur would work. This back-up manoeuvring was an essential and integral part of his overall strategy.

Support for Huxley

Huxley's term "Sauropsida" gained a wide currency, and the avian ancestry became a key element in the palaeontological campaign for evolution. In the sixth and final edition of the *Origin*, Darwin fairly chortled over the "most unexpected manner" in which *Archaeopteryx*, ostriches and *Compsognathus* had helped bridge an embarrassing gap. In Philadelphia, Cope had independently published an extract on "the extinct reptiles which approached the birds" – using dinosaurian stance and the little *Compsognathus* as proof, while, interestingly, shunting *Archaeopteryx* off onto the pterodactyl line. At Yale O. C. Marsh agreed with Huxley that "Birds have come down to us through the Dinosaurs" – appreciating that his own toothed *Hesperornis* (a "carnivorous swimming Ostrich" Marsh called it) rendered a dinosaur-ratite link still more probable. The New World, partly as a result of these dramatic finds and partly because of the dominant social trends, welcomed Huxley's speculations. In Germany, too, the rapid spread of *Darwinismus* helped matters along. Andreas Wagner, who might have proved a formidable opponent, had died shortly after describing *Compsognathus* and *Archaeopteryx*, the very fossils which Darwinian propagandists like Oscar Schmidt at Strasburg considered "unimpeachable" evidence of the avian line of descent.[28]

Of greater importance was the way Huxley carried more cautious palaeontologists at home, while others resisted as a matter of principle – even though neither had much love for Darwinism with its wretched fortuity. Younger Darwinians, like Lankester, were obviously converted; the real challenge lay in wooing the conventional and uncommitted, most of whom accepted a limited descent so long as it retained a design component. And with notable exceptions, Huxley succeeded in swinging the vote. As a result, by 1875 a number of geologists, at best indifferent to Darwin, were nonetheless committed to a dinosaur-bird "affinity".

One such was John Phillips (1800–1874), an eminent, middle-of-the-road geologist. The *Geological Magazine* once quipped that he was "to the *hammer* born", being nephew to the surveyor and founder of English stratigraphy William Smith. Certainly, having worked his way up from the Yorkshire Philosophical Society, via King's College (where he was Lyell's successor) to Buckland's Chair at Oxford, Phillips was pre-eminently placed to test Darwin's theory against the fossil record. Darwin found his *Life on the Earth* (1860) "cautious", but not nearly so "hostile" as he had feared. At times quite the reverse; Phillips even took a crafty swipe, presumably at old Sedgwick, for the "heavy condemnation" usually "heaped" on such speculations. Yet Phillips too was convinced that Darwin had exaggerated the geological gaps, whose existence conveniently explained the 'missing links'. He actually embarrassed Darwin by recalculating geological time, trimming the figure down to 96 millions, thus proving Darwin's estimate of 300 millions since Wealden times alone wildly inaccurate. But what really galled Phillips – like so many of his generation – was the idea of *random* variation, and he asked how, if nature selects only the best, "this beneficent personification [is] to be separated from an ever watchful providence". Darwin lost his patience, preferring to be attacked outright than treated in this "namby-pamby old woman style". "Unreadably dull" was his verdict on *Life on the Earth*, the inevitable penalty for caution at a time of scientific excitement. Nevertheless, Phillips had probably spoken for the silent majority at grass-roots level, those who applauded Darwin's efforts to "tear aside the veil" but found themselves trembling in his wake. Even in 1871 Phillips remained lukewarm towards evolution, with some justification. His study of local Oxford rocks had indeed shown that species succeed one another as if by normal generation, but at an extremely slow pace. At this rate, genera and higher taxa would have required hundreds of millions of years for their emergence, and there was simply not enough time. He therefore proscribed evolution within "narrow [i.e. specific] limits" and provisionally left higher taxa the result of some extraordinary preCambrian event.

Unlike Owen, Phillips was neither standoffish, nor a die-hard. Mantell had once described him as "one of the most pleasant, modest, sensible men I ever met", and in old age he was quite able to "keep pace" with geological events. At Oxford he presided over some of the best dinosaurs in Britain, notably Buckland's local *Megalosaurus*, and it was to the new museum that Huxley was invited in 1867. Not unnaturally, he took the opportunity of talking over mutual problems while admiring the Oxford fossils. Phillips had already recognised that the 'pachydermal' *Megalosaurus* was wrongly built, and together they dismantled the "Great Lizard" and reconstituted the pelvic girdle from misplaced shoulder bones. This gave Huxley another bird–like dinosaur, and Phillips too was partially persuaded, even though the monster was anything but avian in size (twenty-five feet) or weight (two to three tons in his estimation). Phillips was not young; he was 71 at the time of writing his *Geology of Oxford* (1871) and was to die three years later from a nasty fall. Yet he was prepared to go far with Huxley and admit that *Megalosaurus* was

> essentially reptilian; yet not a ground crawler like the alligator, but moving with free steps chiefly, if not solely, on the hind limbs, and claiming a curious analogy, if not some degree of affinity, with the ostrich.[29]

Doubtless by 1871 Phillips had been overtaken by his students; one, William Boyd Dawkins, was employing providential evolution as early as 1865, having encountered problems with rhinoceros specialisation which were "incapable of any other solution" (Chapter 6). Yet despite his opposition to natural selection, Phillips made solid gains in the study of species succession. And his doubts about the evolution of higher taxa did not stop him from recognising the megalosaur's avian affinity, even if this was Huxley's major evidence for the "descent" of an entire class.

Huxley inspired equal confidence in younger palaeontologists, even those reluctant to indulge in theoretical matters. The best example was Henry Woodward (1832–1921), founder-editor of the *Geological Magazine* (f. 1864), the young assistant in the British Museum who had examined *Archaeopteryx* soon after its arrival. Though he was – or was to become – an expert on fossil crustacea, Woodward kept close to the bird question. It is said that he "hesitated to deal with principles" and declined to bring doctrines (like evolution, which he accepted) to problems in palaeontology. True perhaps of his crustacean work, yet he did make a show of supporting Huxley: strengthening the sauropsid ancestry by tying in the bipedal trackways found in the Solnhofen slate (and attributed to *Archaeopteryx* by some) with reports from Australia that the living Frilled Lizard "perches" on stems and habitually runs "on its *hind legs*". Although an officer in Owen's institution, Woodward was loyal to Huxley on the question, having talked it over with him personally. Relations with Owen were perhaps less cordial because the Superintendent was (or seemed) unapproachable. When Woodward obiturised Owen, he was charitable to a degree, acknowledging the man's talents and undoubted greatness; yet he noted a distinct lack of generosity – and that his superior, always suspicious of younger rivals, would conceal choice fossils from their "inquiring eyes".[30] Was Woodward speaking from first-hand experience? If so then any lingering doubts about loyalty would have been crushed, leading to the assistant's desertion to friendlier climes.

The case of John Whittaker Hulke (1830–1895) at the College of Surgeons is again quite different; he was already a renowned ophthalmologist before turning his hand to geology and throwing in his lot with Huxley. Never having come under Owen's jurisdiction, Hulke gave no indication of having suffered the apprentice's fate. His intriguing migration from the human eye to the reptilian retina and thence to *Iguanodon* and its kin was essentially complete by 1868, when he was elected Fellow of the Geological Society. (The move was not as odd as it sounds; Hulke had long been a spare-time fossil prospector, concentrating on the Wealden cliffs of the Isle of Wight.) Entering geology laterally, as it were, at the moment of Huxley's triumph, Hulke proved not only a competent morphologist but a formidable ally. Perhaps even a natural one, despite (or rather because of) his Calvinistic leanings; he was a man of "strict" views and "austerity that amounted to harshness". Of Dutch Reformed descent, he was deeply religious; *The Lancet* commented that "his Protestantism was of the intolerant kind", to the degree that his judgments sometimes "seemed unnecessarily severe". This perhaps complemented Huxley's equally harsh 'scientific Calvinism', which banished "special providences" and left a rigidly-determined universe. Certainly Hulke took an immediate liking to his agnostic counterpart and entered the palaeontological fray on Huxley's side against the Romantic and "liberal" Owen.

Although Hulke was described as "a palaeontologist of no ordinary merit", and less soberly as "one of the first authorities in the world on vertebrate palaeontology", he in fact restricted himself to a tiny corner, concentrating on Isle of Wight dinosaurs. For a time he had access to the best private collection of its kind, belonging to the Rev. W. Fox at Brixton, I.o.W. Although apparently happy to have Owen share the collection, Hulke's first thought on locating a complete *Iguanodon* braincase in September 1869 was to show it to Huxley and take the "greatest pleasure" in asking *him* to describe it.[31] At the time Huxley was overworked and turned it down, but he did keenly assist Hulke at all stages of preparation and description. And on publication he diplomatically "congratulated" the Geological Society on this progress in fathoming the "ornithic" reptiles. Hulke had in fact drawn no pro-Huxley conclusions, but shown this friendship and interest he began intensive study of *Iguanodon*'s ankle and hip bones, publishing a string of papers (1873–6) in the Society's *Quarterly Journal* which endorsed Huxley on every point.

While I am not suggesting that Huxley's personal encouragement was the sole factor in swaying Phillips, Woodward and Hulke, it might have been a key one. Unlike Owen, Huxley was approachable, interested and willing to speculate (which gave his science the tinge of excitement Owen's now lacked). Others, however, were never persuaded – those who had broken off personal relations with Huxley, who were appalled by his ideological crusade or who had philosophical grievances – and it was they who now put up strong opposition to the bird ancestry.

Dissent

> There is perhaps no single point along the whole line of defences raised by the opponents of the theory of Evolution which has been more warmly contested than the . . . relationship of birds to reptiles.
>
> Henry Woodward in 1873.

Woodward probably meant opponents of *Darwinism* rather than "Evolution" per se, since a number of idealists did put forward counter proposals. They clearly sensed that Huxley was scoring points specifically for Darwin, hence they combatted his bird ancestry and looked for a rival which might better serve Owen's providential theory of "derivation". Indeed this opposition is revealing in that it shows a clearly marked pattern. The most prominent dissidents, all of whom refused to be silenced, were the descriptive palaeontologist Harry Seeley, that thorn in Darwin's side, St George Mivart, and of course Owen himself. Each had openly expressed doubts about the dinosaur relationship (or its Darwinian explanation): Seeley denied it immediately, only wavering after two decades, Owen took pains to devise an alternate route, and Mivart (no palaeontologist himself) played up the non-Darwinian consequences of Huxley's position. All three were theist, and although Mivart was a liberal Catholic and Owen a Broad Churchman, both would have agreed with Seeley that "the plans of animal life . . . are the product of divine law".[32] Platonism, or so it seems, was the common bond; each showed Romantic leanings and accepted certain divine ideas behind the diversity of Creation. The Archetypes, as we have seen, were immutable, so that however much Owenism was "refreshed and invigorated" by its evolutionary bath, as Mivart maintained, there was always a certain ambiguity to the cleaned-up product. For instance, evolution had practically no impact on Seeley's science; he hardly ever mentioned the word in almost fifty years of publishing. And while Owen and Mivart embraced a "secondary creational law" (explaining the progressive incarnation of the Archetype by means of an internal, preordinated impulse), they rigorously opposed Darwinian selection.

Dissent thus rested on an ideological foundation; Huxley's bird-ancestry was unacceptable because it was being used indirectly to underpin a Lucretian world-view and because its leading exponent – Huxley himself – was a vehement anti-Platonist. This will become clearer as we examine individual cases.

Seeley perhaps is special. His anarchic tendencies (he surely stood to contest two papers in three at the Geological Society!) set him off; even in his own day he was considered strikingly individualistic. Hence at the conclusion of Huxley's lecture purporting to show "Further Evidence of the Affinity between Dinosaurian Reptiles and Birds" in 1869 he rose as usual to register his dissent, dismissing the long dinosaurian leg as "adaptive", i.e. related to lifestyle rather than reflecting "any actual affinity with birds".[33] It is easy to ignore this as yet another intransigent stand; yet he did have strong, Romantically-inclined reasons for keeping birds allied to pterosaurs, whose structural "plan" he thought they shared. But to do Seeley justice (an unusually difficult task) we must treat him separately (see Chapter 6). Mivart, on the other hand, is far harder to dismiss.

St George Jackson Mivart (1827–1900) was Darwin's most persistent critic; "none proved so formidable", wrote Peter Vorzimmer, who even saw him badgering an irate Darwin into premature retirement (from evolutionary affairs, at least). Mivart was a zoological aristocrat: sauve, well-spoken, Harrow-educated (his father owned the Mivart Hotel, later to become Claridge's). He was also caught in the unique – at first sight impossible – position of being both an admirer of Owen and protégé of Huxley, a situation made more incongruous by his early conversion to Catholicism (1844). Although called to the Bar in Lincoln's Inn in 1851, Mivart never actually practised law, like Seeley preferring the less profitable pursuit of nature (and like Seeley being inspired by Owen's lectures at the College next door). Mivart was introduced to Huxley early in 1859, and like all first-time acquaintances was struck by those "deep set dark eyes". In fact he was perfectly placed to contrast his twin heroes: Huxley, intense, quick-witted, frank, and Owen, plodding and taciturn, perhaps even uncomfortable. But such was their combined stature that a reference from each had been enough to secure for Mivart the Chair of Zoology at St Mary's Hospital in 1862. Even as a professor he continued attending Huxley's lectures at the School of Mines; indeed, they became close friends, dining together and arranging family visits. When the going got rough, Huxley complained that Mivart stuck to him like a "leech",[34] although at the time he was only too pleased to have an admirer and even waived tuition fees. Mivart, then, had been an obedient Darwinian. As he later told Darwin:

> I was a thorough going disciple of the school of Mill, Bain, & H. Spencer & I am so strongly persuaded of the intellectual error & moral mischief of their views only because they were once so completely mine – As to "natural selection" I accepted it completely and in fact my doubts & difficulties were first excited by attending Prof. Huxley's lectures at the school of mines.[35]

Huxley's ideological bent, his natural antipathy towards Rome and growing mechanist outlook must have subtly worked his young Catholic disciple to crisis point. Racked with doubts and unable to condone this illegitimate use of evolution, Mivart began showing signs of distress. To Darwin, for example, he wrote:

> For my part I shall never feel anything but gratitude & sincere esteem for the author of "natural selection" but I heartily execrate some who made use of that theory simply as a weapon of offence against higher interests and as a means of impeding Man's advance towards his "end" whatever may have been his "origin".
>
> In my wandering about Italy I have been amazed and saddened to see our friend Huxley's "Man's place in nature" for sale at most of the railway stations amongst a crowd of *obscenities*.[36]

What completed Mivart's estrangement was probably the efforts of younger Darwinians to finish Huxley's work and eradicate all traces of Platonism. In 1870, shortly after coming down from Oxford, Ray Lankester published a critical paper "On the Use of the Term Homology in Modern Zoology" in the *Annals and Magazine of Natural History*. He advocated purging words like "homology" which belonged "to the old Platonic school". Homologous structures – for example the human arm, whale's fin, and bat's wing – were only related in an *ideal* sense, i.e. because they adhered to the same archetypal plan. According to Lankester, "homology" was meaningless without Owen's "ideal type" as a standard. A new word was needed which could be "defined afresh in accordance with the doctrine of descent", and he proposed "homogeny", meaning similarity as a result of shared ancestry. This was doubly offensive to idealists, many of whom saw "homology" as Owen's crowning achievement and an explanatory end in itself. Darwin on the other hand publicly praised Lankester's "remarkable paper" and was convinced that Huxley's scientific 'son' was destined to "become our first star in Natural History".[37] (He might have, but for a bellicose disposition.) Mivart was angry and answered back. He rebutted Lankester's proposals, and in the *Annals* vindicated Owen, for whose "self-consistent theory of the vertebrate skeleton" science was so "deeply indebted". What was needed, he urged in 1870, was not relegation, but reconciliation between Platonism and Darwinism,

> for much as G. St-Hilaire and Cuvier contended together in France, so in England two schools have sought for the separate and exclusive establishment of views really capable of a fruitful and harmonious coexistence.

He recognised that Owen's "ideal archetypes" appeared "fundamentally baseless" to Huxley, and that natural selection was thought to give "the '*coup de grâce*' to such fancies". Yet a higher teleology was compatible with evolution. Even "the special Darwinian form of it" did not rule out the possibility of creation "according to certain ideal types". More and more the idealist, Mivart thought Lankester's "question as to the *mode in which* such homologies arose" of secondary importance.[38] The morphological goal, irrespective of how it was attained, was the key issue; in fact, so many species were convergent in appearance and yet unrelated that Mivart dissociated homology or structural similarity from the idea of common origin. Homology retained its cardinal importance, against which Lankester's "homogeny" paled into insignificance.

But anti-Platonic feelings were running high in Darwinian circles and Mivart was seen only to condemn himself. When he published some "Difficulties of the Theory of Natural Selection" in the Catholic *Month*, Huxley's colleague Flower took up the challenge. Newly instated in Owen's old Chair at the College of Surgeons, Flower delivered an inaugural lecture (1870) which indicted Owen and vindicated the Darwinian system. He denied that natural selection was incapable of explaining the close similarity of placental dog and marsupial wolf, as Mivart claimed. Quite the reverse, this convergence is what we might have expected given similar selection pressures. Mivart's case was tossed out and Flower denounced transcendentalism for succeeding only in burying anatomy under a plethora of incomprehensible names. Like Lankester, he urged a thorough revision of textbooks and teaching methods to usher in an "epoch of revolution . . . in our Profession".[39]

Mivart was beleaguered. Even before publication of his major work *On the Genesis of Species* in January 1871, Huxley, Lankester, and Flower had all come out against him. The *Genesis* electrified the situation. Darwin thought it grossly unfair and largely the product of "bigotry". Their letters became fraught, a desperation of pleasantries, denials of bad intent, yet mutual dissent. It took exactly a year for relations to collapse, a year which saw Mivart's hostile *Quarterly* review of Darwin's *Descent of Man* countered by Huxley's stinging rejoinder. In January 1872 Darwin dropped Mivart a note saying he would no longer read his papers,[40] and Huxley rather callously told Mivart to stop "running with the hare and hunting with the hounds".

If the *Genesis* was Mivart's attempt to undo the ideological damage wrought by his erstwhile colleagues, it also shows him wholly disillusioned with the Darwinian mechanism. It was designed to render natural selection impotent, or at least subservient to an inner driving mechanism. To the extent that Pius IX awarded him an honorary Ph.D. in 1876 he succeeded; but then the Pope was barely infallible and Rome still marginally liberal (Mivart's fortunes at the hands of the Roman curia were to take a markedly downward turn in later years). The *Genesis* acted as a logical framework to hang the diverse strands of Darwinian opposition. The *Quarterly Review*, rabid at the prospect of chance creeping in, had already asked how even an infinite number of accidents could conspire to transmute the entire mosaic of bodily tissues, while keeping them delicately synchronised. Chance was preposterous, "there must be a guiding principle within variability itself," pushing life forward in a well-planned drive. This was Mivart's launch point. The innate drive he saw manifesting in convergent structures, and in "identical" yet unrelated animals. Darwin's single tree was ousted by a "grove of trees", each rooted separately but with branches intertwining and stretching upwards together. The new fossil work provided a perfect example, and Mivart carefully exploited Huxley's apparent confusion over dinosaurs, *Archaeopteryx* and ostriches. Had flying birds descended from pterosaurs and ostriches from dinosaurs, no amount of fortuitous variation could explain the marvellous confluence between the two avian types today. Conversely, if *Archaeopteryx* was descended from giant struthious birds, could random variations explain its similarity to pterosaurs? – a similarity which extended to brain shape, as Seeley had shown. Birds had grown to resemble pterosaurs because the process has taken place not by merely haphazard, indefinite variations in all directions, but by the concurrence of some other and external natural law or laws co-operating with external influences . . .[41]

This "innate power" could channel the development of unrelated organisms along parallel or even convergent pathways. All vertebrates which flew would therefore tend towards an optimum aerodynamic shape.

Huxley came crashing down on Mivart in a story that is only too well known. Mivart's specific avian criticisms – in fact almost all his scientific criticisms – went by the board. What Huxley really loved were metaphysical and military issues, and the *Genesis* afforded him an unbridled opportunity to pitch into "that vigorous and consistent enemy of the highest intellectual, moral, and social life of mankind – the Catholic Church." But for all the exegetics and quibbles over whether or not the Spanish Jesuit Suarez provided for "derivative creation", Mivart was really being pinned out, as Huxley admitted in a candid letter to Hooker, for being "insolent to Darwin".[42] The young guard smartly closed ranks to exclude the heretic and Hooker was still prepared to blackball Mivart's application to join the Athenaeum Club years later. The punishment would have been out of all proportion had the crime been solely scientific. But Mivart's untimely and unexpected pledge of support for the despised Platonism stabbed at the very heart of the new movement; a fact made worse by his almost becoming one of Darwin's inner circle. His precipitous desertion called for a show of strength, as much to warn the faithful as frighten the offender. Mivart thus found himself excommunicated by bell, book (*The Origin*) and candle, as he was later to be by the Church itself.

An outcast, he not unnaturally turned to Owen for help. He knew Owen would be "amused" to hear that Huxley was berating him (Mivart) "for not showing sufficient reverence for Mr Darwin". The irony, of course, was that Huxley himself "some years ago very vehemently repudiated the notion of *reverence for Authority* in matters of science", and Mivart asked Owen's "confidential" help in illustrating this Janus-faced behaviour to strengthen his rejoinder. He also hoped to have "a little friendly chat",[43] but the outcome is unknown. Owen might have been ice-cool – after all,

Mivart had been treading Huxley's path for a decade. To suddenly find the bulldog biting him and expect Owen's sympathy was a bit much. Still, this is the first hint that, by the early seventies, Platonists were being thrown together for mutual protection.

Palaeontology of Despair

> . . . science will accept the view of the Dodo as a degenerate Dove rather than as an advanced Dinothere.
>
> Owen in 1875, dispensing
> justice to Huxley and Darwin alike.

The period 1868–1874 was thus a harrowing one for Platonists. Militancy among materialists had shown an unprecedented upsurge, bringing with it a hatred of all things Romantic. The new middle-class professionals had begun to seize the organs of scientific power in the capital. With extremists in control (or so he thought), Mivart no longer felt able to support the party, and he seceded in 1869, quietly followed by a number of influential palaeontologists (Chapter 6).

Inevitably, the main strength of the Darwinians lay in their social cohesion. This tight-grouping was evident in the almost simultaneous assaults on Mivart. The renegade stood flanked on all sides: Lankester expunging his ideal terminology, Flower tackling his scientific criticisms, and Huxley demolishing all prospects for a Catholic evolutionism. The Darwinians bore down on Mivart, attempting to isolate and prevent the spread of such disaffection. This unmistakable social network is something Platonists lacked. While not scarce, they were unorganised and socially fragmented. Darwinism, by contrast, was perfectly tailored to fit over the middle-class professional network which had sprung up in the fifties. Thus its social base was ready-made, and the professionals to whom it appealed were by 1859 already sliding into positions of power. This trend continued as Huxley's students captured key academic posts in the sixties and seventies. Like Flower, eased into the Royal Society under Huxley's auspices, and groomed as his successor in the Hunterian Chair, Lankester had risen through Huxley's intercession. Not long after his pro-Huxley paper in the *Annals* a Fellowship in the sciences was created at Exeter College, Oxford, and Huxley effectively secured it for him. And when Lankester then complained of the "medieval folly and corporation-jealousy and effete restrictions"[44] of the place, he was relocated at University College London, where the Chair of Zoology had fallen vacant on Grant's death.

Some of the support Platonists might have summoned had been jeopardised by Owen's own intransigence. Huxley, Flower, and Rolleston had capitalised on this in their effort to disgrace him during the ape-brain debate. Rightly or wrongly, Owen had been branded a liar, and his name dragged through the mud (or at least the medical press and popular weeklies). No doubt it was mostly his fault, yet the treatment was particularly harsh, and for Owen, deeply distressing; he was never to forgive or forget, his anger and frustration spreading through his science like a cancer. *Ten years* after the event, in an outspoken letter to Tyndall, he gave vent to his feelings. Tyndall, who had written secretly in June 1871 imploring Owen to heal the breach with Huxley, received this tart reply:

> Prof. Huxley disgraced the discussions by which scientific differences of opinion are rectified by imputing falsehood in a matter in which he differed from me. Until he retracts this imputation as publicly as he made it I must continue to believe that, in making it, he was merely imputing his own (base and mischievous) nature.[45]

Given the circumstances, it was foolish to have imagined anything different; their paths had grown so divergent, their ideologies at such loggerheads, and their science so riddled by cat-and-mouse games, that matters had simply worsened. "I had hoped against hope", Tyndall sighed, "for it has saddened me beyond measure to witness the conflict and consequent estrangement between two such men as you and Huxley." Not that Huxley would have welcomed any easing of relations (and how odd that Tyndall was blind to the fact). With Owen the Darwinian *bête noire* and the butt of Huxley's latest palaeontological campaign, reconciliation might have been disastrous.

By 1871 Huxley had only just completed demolishing Owen's

majestic dinosaur, leaving mopping up operations to Hulke and Woodward. Even as he replied to Tyndall, the Grand Old Man of palaeontology was being pushed into a corner. Perhaps Owen should have stood his ground, but he was alone and retreated. He now swung the full might of his palaeontology into reaction. Naturally he rejected Huxley's bird ancestry outright, and intended his new palaeontological creation to undermine not only Huxley's avian work but its Darwinian wellsprings. In short, from 1872–8 Owen refashioned the dinosaur, undoing the marvellous work he had done in the forties, deliberately making the beast inaccessible to Huxley, Darwin and their supporters. He was so blatant about it as to leave no other explanation possible.

Huxley's final provocation might have been the last straw. In 1869 he recruited a new Isle of Wight dinosaur *Hypsilophodon* for the cause, assuring the Geological Society that it "affords unequivocal evidence of a further step towards the bird"; and the claim now sounded less outrageous given the reptile's turkey-size. Pushed to the fore and made an evolutionary test case, the little *Hypsilophodon* began to attract supporters. It became 'hot property', with Huxley, Owen and Hulke all working on it simultaneously. Hulke probably switched his attention because of this supposed avian affinity, and he backed Huxley to the hilt – deliberately clashing with Owen, who regarded the new fossil as merely a dwarf species of *Iguanodon*. Meanwhile, Huxley began laying siege to Owen's 'pachydermal' monstrosity, using tactics familiar to anyone who remembered the ape debate. He accused Owen of filching his 'pachydermal' model from von Meyer – who had labelled iguanodons and megalosaurs *Pachypoda*. Even Owen's replacement definition was botched, and he failed to question the basic assumption that dinosaurs "made the nearest approach to mammals". From Huxley's tone, it sounded as though Owen had compiled a catalogue of glorious errors:

> Every character which is here added to Von Meyer's diagnosis and description of his *Pachypoda* has failed to stand the test of critical investigation; while it is to birds and not to mammals that the *Dinosauria* approach so closely. There is, in fact, not a single specially mammalian feature in their whole organization.[46]

In the end Huxley allowed "justice to give way to expediency" and he retained the name Dinosauria for the huge *Iguanodon* and *Megalosaurus*; in which case it was wholly inappropriate for the tiny, delicate *Compsognathus*, and Huxley separated this off into a parallel group of equivalent status, the Compsognatha. And in an effort to carry his ideas through into the classification, he united them all under the banner Ornithoscelida (i.e. "birdleg"). Owen was outflanked. By 1870 Huxley had challenged conventional wisdom on dinosaurian physiology (indicating an avian anatomy and even "hot blood"), given the reptiles a bird-like bipedal stance, and finally changed the classification to rubber-stamp the "ornithic" product.

Owen's integrity had again been impugned; he stood accused of plagiarism – practically of ineptitude in carrying out his dirty deed. To this shabby treatment he responded predictably. He denounced the dinosaur-bird ancestry as another of Huxley's "sensational tricks" and staked out a rival position in his Palaeontographical Society memoirs. But to dodge Huxley's obstacles he had to steer his dinosaur on a crooked course. This he did with a certain precision, even self-consistence, but still the put-upon beast emerged looking the worse for wear.

Ostriches were not the progeny of dinosaurs. Owen saw them as a classic case of "Buffon's belief in the origin of species by way of degeneration". Some time in the past, perhaps on an island, with plentiful food and a lack of predators, birds had descended from the trees and their wings atrophied. But if right-minded biologists accepted "the Dodo as a degenerate Dove" then "whence the dove?" Not from dust, and certainly not from Prof. Huxley's dinosaurs. Owen hoped to be spared "the bottomless pit of Atheism" for suggesting an alternative natural route. Admittedly pterodactyls differed from birds in possessing teeth, tails and an elongated wing finger, but on two counts *Archaeopteryx* bridged the gap. Like Cope and Seeley, Owen placed *Archaeopteryx* between birds and pterosaurs, although "how Ramphorynchus [sic] became transmuted into the Archaeopteryx" was an open question.[47] Owen took a functional approach to the rest of the problem. The "abettors and acceptors" of natural selection had failed miserably to explain why the fore-limbs had shrunk in some dinosaurs. It was insanity to imagine them walking daintily on their hind legs like Phillips's ostrich; Owen delved deeply into joints, bones and muscles to prove that the "gravigrade" monsters could no more rear up than fly. His real functionalist explanation, though, was in some senses disappointing. He noted that extinct aquatic crocodiles had small fore-limbs for streamlining. With the tail providing propulsion, the arms were tucked against the flank "in the forward dash" of the predator. Owen now took a giant leap backwards: he proposed that *Iguanodon*'s tail was similarly compressed to act as a propulsive caudal fin, adding that the degree of fore-limb reduction makes a convenient gauge of the dinosaur's aquatic adaptation. The idea was not ludicrous, in fact enormous sculpted vertebrae, reminiscent of a whale's but actually belonging to the "whale saurian" *Cetiosaurus*, seemed to point in the same direction. Yet however good the evidence, it was impossible not to imagine Owen up against a wall. He bravely fought a rearguard action, but his splendid '41 'pachyderm', already crippled by Huxley's shafts, was forced to flee into water and swim away.

5

"Phylogeny"

England's Lack of Response?

It seems rather odd that the publication of Darwin's book did not at once cause a rush among palaeontologists to find more evidence as to the manner in which new fossil species appeared in the rocks. One might have thought that attempts in evolutional palaeontology would have been very much to the fore in the eighteen-sixties, but it was not till nearly the end of the century that deliberate studies began to be made in Britain.

John Challinor in his bibliographic *History of British Geology*.[1]

Surprisingly little genealogical research was carried out in mid-Victorian England. Disregarding Huxley (which Challinor does), there was nobody to rival the respectable pedigree work emanating from the Continent or America,[2] let alone the speculative exercises of the German, Ernst Haeckel (1834–1919). Ironically, one of the most challenging projects in England, Phillips' study of Jurassic invertebrate succession,[3] was literally that: an attempt to *challenge* Darwin's belief that evolution could work above the generic level. Phillips found no evidence for the gradual emergence of new orders or classes.

So why did the *Origin* fail to make a greater impact on British palaeontologists? The social historian David Elliston Allen puts the lack of enthusiasm down to a growing professionalism among geologists; having defined a narrower set of problems, to be solved within their disciplinary boundary, they were reluctant to resort to grandiose multi-disciplinary solutions, of the sort offered by Darwin.[4] Obviously not all palaeontologists were oblivious to the *Origin*. Huxley was devoting increasing time to the subject; to leave him out of the picture is – as Cope said of Darwin's failure to recognise the progressive element in evolution – to leave the Prince out of *Hamlet*. But the fact remains that, apart from Hulke, Huxley was almost alone in tackling genealogies. My object in this chapter is to explain why this should have been so, and, in particular, to investigate the *social* reasons why Huxley more than any contemporary palaeontologist was eager and willing to import the new German approaches to fossil history.

Other explanations of geology's neglect of Darwin spring to mind. We might, for example, see the *Origin* as primarily a sophisticated exercise in biogeography, and Darwin's reconstruction of variation, migration, and isolation as a neat solution to the difficulties he encountered with the distribution of, say, Galapagos and South American finches. Not surprisingly, his most ardent supporters, Hooker at Kew and the co-founder of natural selection, Alfred Russel Wallace, were biogeographers, not palaeontologists; and it was to geographical distribution with its perennially knotty problems that the *Origin* obviously rendered greatest service. Looking deeper, this emphasis itself reflects England's maritime tradition and expanding global empire. Darwin, Hooker, and Huxley had all spent protracted periods aboard Royal Naval surveying vessels, which were engaged in oceanographic measurements (ultimately with a view to securing British trade routes). Huxley actually met his future wife in Sydney. And Wallace was a kind of global vagrant, who travelled extensively in the jungles of South America and the Far East, where he finally conceived natural selection.

The crucial importance of biogeography for the inception of Darwin's ideas is reflected in the rapidly expanding literature on the subject, especially in the *Journal of the History of Biology*. But whereas geographical distribution had a venerable history and was anchored in the Lyellian science imbibed by Darwin, the concept of "phylogeny" (i.e. the study of the evolutionary routes followed by particular organisms) was largely a sixties' invention – born of Haeckel out of Darwin, one might say. But why? Why did it take a *German* to draw the phylogenetic implications out of Darwinism?

For one thing, in Germany at the time there was less emphasis on global distribution patterns. Indeed, the German states before unification (1866–1871) had little maritime experience, and Haeckel in 1865 actually had to ask Huxley to get him onto a *British* surveying ship (although Huxley counselled him to "stop at home, and as Goethe says, find your America here").[5] Not that Germans failed to voyage to distant continents; one has only to think of Humboldt's travels, and his best-selling accounts. The point is that Germany lacked a colonial empire, something Haeckel was to regret on witnessing the French excursion into Algeria. No overseas colonies and no protected trade routes meant for the most part no continuous supplies of fossils or exotic cadavers, as were being shipped to Paris and London from the four corners. Hence German science was orientated less towards biogeography than *Naturphilosophie*, with its paradigm in embryological growth; and it was from a Romantic seed planted in Darwinian soil that Haeckel's famous (or infamous) family trees sprang.

Geographical distribution was also attractive in Britain because it was largely empirical; it fitted in superbly with the methodological make-up of early Victorian science. The Darwinism which grew from it was another matter: largely hypothetical, occasionally grandiose, and to many morally questionable. Thus anyone who wished to use it to build family trees had to forestall the expected methodological criticism. Sedgwick and Phillips had insisted that geology should be descriptive, factual, and starting in "that tram-road of all solid physical truth – the true method of induction". Accordingly the only valid "generalisations" were those arising almost unaided out of the piles of dry facts. Darwin himself attempted to avert criticism on this score by opening his *Origin* with the disarming claim that the book rested on twenty years' worth of patient fact finding. Still Sedgwick accused him of abusing the inductive method and starting off in "machinery" as wild "as Bishop Wilkins's locomotive that was to sail with us to the moon." Geology had a fierce pride in its untainted inductivist base (cultivated in the thirties partly to disbar the lunatic fringe of Scriptural cosmogonists; indeed, crude reconciliations were still slightly irritating, judging by the fact that the new middle-brow *Geological Magazine* in 1865 devoted three pages to slating the Rev. Thomas Marsden's *Sacred Steps of Creation*). At the learned societies this descriptive bias was evidently responsible for the shunning of any

discussion of Darwinism at meetings. Older geologists still be-
laboured the point: Phillips in his 1860 address to the Geological
Society boasted of "our emancipation from the tyranny of hypothesis", and looking over the year's productions in the *Quarterly
Journal* praised the "practical character of the papers".[6] Given this
methodological bent, it is not surprising that palaeontology continued producing some of the most doggedly descriptive practitioners, even in late Victorian times: Harry Seeley (Sedgwick's
one-time assistant) perhaps being the most defiant – although I shall
argue in the next chapter that a fine study of his science still reveals
strong philosophic assumptions and a recognisable social commitment.

On the other side, the bourgeois invasion of science in the 1850s
and 1860s quickly undermined Sedgwick's position. New-breed
scientists with an evolutionary axe to grind, those who shaped and
used Darwinism to further their social interests (notably Huxley,
Tyndall, and the *x* Club members), were bound to deplore "the
stream of cold water which has steadily flowed over geological
speculations", as Huxley said in 1869.[7] To legitimise Darwinism,
they had first to lift the methodological restraints imposed by the
Oxbridge dons. Tyndall of course turned evolution into a glorious-
ly deterministic cosmological system and proclaimed us all souls of
fire and children of the sun. Addressing the British Association in
1870 "On the Scientific Use of the Imagination", he despaired of
the Tories "in science who regard imagination as a faculty to be
feared and avoided rather than employed." What sounded like an
impartial plea for scientific freedom was more obviously an attempt
to lift sanctions against the new naturalism, enabling him in the
notorious Belfast address (1874) to "wrest" from theology "the
entire domain of cosmological theory". And since he sought at least
partial justification in the *Origin*, he was earnest in his hands-off
demand, warning fellow scientists "to be cautious in limiting [Mr
Darwin's] intellectual horizon."

In this chapter, then, I shall briefly discuss the generation and
meaning of Haeckel's "phylogeny" in Bismarck's Germany, and
argue that Huxley in England, leading the bourgeois faction most in
need of a 'pedigree', was best prepared to receive the new Haeckelian approach. To appreciate why necessitates our looking at Huxley's social position in what he once called "my scientific young
England" – with the clear emphasis on "my". This means looking

beyond his immediate *use* of fossils, e.g. his manufacture of bird-dinosaur links to undermine Owen's edifice, and concentrating on the more general shift from a world of geologic 'persistence' to one of limited progression. From this it will be apparent that the concept of "phylogeny" boosted Huxley's political power, in terms of his middle-class struggle, and thus it was an inextricable part of the bourgeois takeover of London science.

Haeckel: In Support of Science and State

The character of Haeckel was forged amid circumstances that have largely passed away from the scientific world of our time. The features even of the world he has worked in of recent years in Germany are so different from our own that no Englishman can understand him without sober study of his life.

> Joseph McCabe in the Introduction to his translation (1909) of Bölsche's biography, *Haeckel: His Life and Work*.[8]

Erik Nordenskiöld acknowledged Haeckel's *History of Creation* (1868) as "perhaps the chief source of the world's knowledge of Darwinism". Exaggeration or no, the popular appeal of the man claimed by Büchner to be the "German Darwin" was phenomenal. His *Riddle of the Universe* (1899) sold 100,000 copies within a year in Germany alone, and in Britain the sixpenny Rationalist Press edition did equally well. However only in a narrow sense was Haeckel the "German Darwin", and the title probably hides more than it reveals. Haeckel himself admitted that Darwinism was "only a small fragment of a far more comprehensive doctrine – a part of the universal Theory of Development, which embraces in its vast range the whole domain of human knowledge."[9] Also, in so many ways Darwinism was peculiarly English: in its debt to Paleyite natural theology and Malthusian population studies, in its subtle recasting of Sedgwickian geology and emphasis on fossil and recent distribution – even, Marx laughed, in Darwin's "crude" prose style. Social conditions in Germany in the 1860s were profoundly different, and not surprisingly Haeckel acted as a kind of distorting lens, adapting the *Origin* to nationalist needs. This necessarily entailed its *r*ecreation into a culturally-accessible form, in the process of which he produced some essentially novel concepts in palaeontology. It is no coincidence that as Germany reunified under Bismarck, "the practical creator of history",[10] so Haeckel developed the concept of *phylogeny*, or racial history of the *phylum* (both words of Haeckel's invention). To go with the new historical sense, he had presented a palaeontological parallel – the pedigree.

The emotional component of Haeckel's *Darwinismus* probably arose from his Free Evangelical upbringing. The son of Prussian Civil Servants, he remained a pious Protestant throughout his university days at Würzburg, at which time he could still condemn the materialist outpourings of Carl Vogt. Evangelism coupled with Haeckel's love of *Naturphilosophie* later led to a kind of mystical nature-worship (again distinguishing him sharply from English Darwinians). Goethe and the influential nature-philosopher Lorenz Oken he declared to be prophetically inspired, and Goethe actually shared the title page of the *Generelle Morphologie* in 1866 with Lamarck and Darwin. About the time Darwin published the *Origin* Haeckel settled in Jena, a university town which held tremendous attraction for him. At one time or other it had welcomed Goethe, Hegel, Oken and Schelling, the Prince of Romantics. Perhaps most of all he valued it because it allowed him to work closely with his friend and mentor, Carl Gegenbaur (1826–1903), under whom he hastily completed a thesis on the planktonic radiolaria in 1861. By 1862 Haeckel was enjoying equal status as Extraordinary Professor of Zoology. It was an exciting time, Haeckel recalled, when the "mystic veil of the miraculous and supernatural" was being lifted by Darwin.[11] He stayed at Jena for the rest of his life, turning down offers of lucrative posts elsewhere.

Jena was a Protestant university and Gegenbaur a Catholic, and relations between the men were understandably to cool as Haeckel's vitriolic campaign against Rome escalated (mirroring the State's own hostility after the 1870 declaration of Papal Infallibility, when Bismarck brought in his restrictive laws against Catholics, banning the Jesuits, etc.). Despite this, in the 1860s Gegenbaur had attempted to bring morphology into line with descent, and Haeckel's conception of hypothetical ancestral vertebrates probably owes much to him. Gegenbaur built up a picture of the primeval ancestors by a kind of morphological extrapolation, the process Owen used to derive the Archetype. Only Gegenbaur's ancestors had a real, not an ideal, existence. For instance, he imagined the shark cranium as the primordial type, from which all other vertebral skulls might be derived. And he envisaged an "archipterygium" or ancestral vertebrate extremity, from which the fish's fin and tetrapod limb might be derived. In an identical way, Haeckel employed hypothetical forms – like the famous "promammal", or first mammal, from which all others sprang (a spectral creature proving very useful to Huxley – see next chapter).

After the tragic death of his young wife, Haeckel attempted to drown his sorrows by throwing himself into the massive two-volume *Generelle Morphologie*, astonishingly written and printed within a year. With the more popular *History of Creation* it formed an extended argument for descent, backed by speculative exercises in genealogy and ancestor reconstruction. He brashly tackled the ape-ancestry of man in a way that made *Man's Place* seem a model of caution, and Darwin admitted that had he read Haeckel earlier he might never have completed his *Descent of Man*. But, unlike Darwin's books, Haeckel's had an overtly political function; they were organs of social reform, and designed to make the new specialist knowledge "fruitful to the mass".[12] In a crudely effective way, for example, he emphasised our ape ancestry because it was distasteful to the vaunted aristocracy – noblemen and dogs, he once remarked, were all of a kind in the womb.

The last decade has witnessed a historiographic overhaul of Haeckel. From being the tireless champion of liberal democracy he has become the more sinister scientific father of proto-Nazism, as portrayed in Daniel Gasman's *Scientific Origins of National Socialism* (1971). Haeckel's early liberalism was of a distinctly German kind. Where British middle-class liberals inherited a strong state and saw their ends best served by *reducing* government interference (finding a champion in Spencer), Germany after unification had a more feeble State apparatus, and the out-numbered middle classes saw their interests best served by strengthening it. Hence they opposed Spencer's "exaggerated self-seeking individualism".[13] Of course, Haeckel's incessant baiting of Church and nobility did attract Left-leaning intellectuals. The cellular pathologist and public health campaigner Rudolf Virchow – in an effort to halt the spread of Haeckel's social philosophy in schools – actually damned it as a "dangerous doctrine" that would "lead to socialism". But the truth is that as Bismarck moved slowly to the Right throughout the century, Haeckel was never far behind.

Gasman suggests that Haeckel underwent almost instantaneous conversion to Darwinism in 1860 for patriotic reasons. Political unification under Bismark raised hopes of a new teutonic superiority, and in Darwin's daring synthesis, Haeckel recognised the spiritual salvation of the new Germany. His "intensely mystical and romantic nationalism", his almost Messianic appeal to the ideal of the German *Volk*, the spirit binding all to the Fatherland – these factors profoundly affected his rendering of Darwinism. The new sense of racial identity at the time of unification was given specific evolutionary form in *Generelle Morphologie*. Welcoming Bismarck to Jena, Haeckel declared: "While the booming of guns at the Battle of Königgrätz in 1866 announced the demise of the old Federal German Diet and the beginning of a splendid period in the history of the German Reich, here in Jena the history of the phylum was born."[14] How are we to read this? An almost spiritual parallel between the superior German brotherhood and the palaeontological (= racial) phylum, progressing through competitive racial 'weeding'? We might be using too much hindsight; nonetheless Haeckel saw racial superiority triumphing in the Western extermination of "lowly" savages, and therefore progression, both of the phylum and Industrial nations, was a deduction from natural selection (as he read it). For Germany to take its rightful place as a European leader, Haeckel now pressed for colonial expansion and commercial exploitation.[15] One begins to understand why he should spontaneously propose Bismarck for an unheard-of honorary Doctorate in Phylogeny. In his religion of evolution, the State's most important intellectual crutch, the *phylum*, was as mystically real as the Volkish bond.

Generelle Morphologie was at once impressive and infuriating. Darwin had been sent proofs and December 1866 found him "groaning and swearing at each sentence", complaining of its unconquerable size and incomprehensible language. (Haeckel admitted that the book sold badly for these reasons, hence his popular version, *The History of Creation*.) Even Huxley with his fluent German found it "an uncommonly hard book", although he seems to have taken vicarious satisfaction in its "polemic *excursus*", telling Haeckel "I have done too much of the same sort of thing not to sympathise entirely with you; and I am much inclined to think that it is a good thing for a man, once at any rate in his life, to perform a public war-dance against all sorts of humbug and imposture".[16] Both of the books alienated and embittered conservatives. True, Haeckel's language was less scurrilous than, say, Carl Vogt's, who occasionally dragged discussion of theology to gutter level (e.g. in designating certain "simious" skulls from the Dark Ages as "Apostle skulls" on the presumption of their belonging to degraded Christian missionaries). But even Haeckel could point to Papal history and the "pious inquisition" as proof that the universe lacked moral order. However, as an "attempt to systematise biology as a whole" along Darwinian lines, Huxley was suitably impressed – according to McCabe, he actually judged the *Morphologie* "one of the greatest scientific works ever published".[17] And both he and Darwin toyed with the idea of a translation for over two years. Savage cuts were needed to avoid "English propriety" taking fright; Darwin even offered to defray some of the cost, and the Ray Society was contacted *à propos* an abridgement, with the "aggressive heterodoxy" toned down. Evidently they met insuperable problems and the scheme fell through. The book was just "too profound and too long",[18] and as a result it is doubtful if Darwin ever read more than a smattering of the German pages.

Friendly links were quickly established. Haeckel called Huxley "the most eminent English zoologist", and he in turn hailed Haeckel as a Coryphaeus among German naturalists. As early as 1865 Huxley had pointed him out to Darwin as "one of the ablest of the younger zoologists of Germany". In October 1866 Haeckel actually spent a day at Down, and Darwin admitted having "seldom seen a more pleasant, cordial, and frank man".[19] There was, perhaps, a certain misfortune in Huxley's christening the infamous ennucleate primal-slime 'organism' from the Atlantic mud *Bathybius Haeckelii*, since it shortly proved no such thing; but as an act of scientific homage it was indicative. And in an *Academy* review the same year, 1868, he paid Haeckel the ultimate back-handed compliment, admitting that *Generelle Morphologie* had "all the force, suggestiveness and, what I may term the systematising power, of Oken, without his extravagance." Despite rounding off his review by endorsing "the general tenor and spirit" of *The History of*

Creation, Huxley felt some trepidation at its apparent mechanistic excesses. He reminded Haeckel that a clockwork universe was ultimately indistinguishable from a teleological one, taking the Tyndallian line "that the existing world lay, potentially, in the cosmic vapour; and that a sufficient intelligence could, from a knowledge of the properties of the molecules of that vapour, have predicted, say the state of the Fauna of Britain in 1869 . . ." The same year he posted to Jena a copy of his Edinburgh lecture "On the Physical Basis of Life", with its criticism of philosophical material-ism "which I should like you to consider".[20] To employ materialist *terminology* was acceptable, even desirable, but to profess know-ledge of ultimate matter (or spirit) was nonsense. However, it became increasingly apparent that Haeckel's philosophy was a world removed from the simple mechanistic materialism that Huxley supposed. While apparently championing a cosmic mechanism, Haeckel talked in ecstatic terms of all nature being "alive", and was soon to dispense to each atom an individual soul; the holy of holies within the cosmic tabernacle which was to form the basis of his religion of nature-worship. None of this would Huxley have understood or appreciated.

But the early Haeckel did influence Huxley, and profoundly. I suspect that Huxley was actually unaware of the strength of a phylogenetic programme until he read *Generelle Morphologie*. Hav-ing proof-read Spencer's *First Principles* he knew that evolution was ripe for a speculative cosmic gloss. But Spencer had done nothing for phylogeny. If anything, his views had been subverted by Huxley, whose 'persistence' caused him to downplay all notion of fossil progress. Now 'persistence' itself was on the wane, and Haeckel's gnarled family trees began to look tempting. Of course there was consternation at his rather rash manufacture of ancestors from embryonic models. More so perhaps than at his hypothetical ape-man *Pithecanthropus*. But Haeckel himself recognised that these "hypothetical genealogies" were nothing more than an approxima-tion, drawn up in spite of the "extreme difficulty of the task" and "discouraging obstacles"; and he hoped that more accurate replace-ments would soon be forthcoming, even if we could never arrive at a "complete pedigree". Darwin had mixed feelings:

> Your chapters on the affinities and genealogy of the animal kingdom strike me as admirable and full of original thought.

Your boldness, however, sometimes makes me tremble, but as Huxley remarked, some one must be bold enough to make a beginning in drawing up tables of descent.[21]

On occasions, historians have been inordinately critical of Haeckel for his rampant phylogenising, and in as much as he resorted to questionable techniques (or even guesswork), the criticism was probably justified. But this is to miss the heuristic function of his trees. As Oscar Schmidt of Strasburg conceded in 1873, "It matters nothing that he has repeatedly been obliged to correct himself, or that others have frequently corrected him; the influence of these pedigrees on the progress of the zoology of Descent is manifest to all who survey the field of science . . ." One only really begins to appreciate this on observing the effect they had on Huxley. Was he signalling conversion when he confessed in his *Academy* review that:

> In Professor Haeckel's speculations on Phylogeny, or the genealogy of animal forms, there is much that is profoundly interesting . . . Whether one agrees or disagrees with him, one feels that he has forced the mind into lines of thought in which it is more profitable to go wrong than to stand still.[22]

Anyway, Huxley's first use of trees immediately followed his reading of Haeckel. He received his copy of *Generelle Morphologie* in November 1866, although lecturing prevented anything but a skim, and the following May he apologised to Haeckel that he had still "not been able to read it as I feel I ought". However, the need for an English abridgement eventually sent him deeply into the tome, shortly after which he constructed his first pedigree. Addressing the Zoological Society in 1868, he suggested that "all the Gallo-columbine birds [partridges and pigeons] must be regarded as descendants of a single primitive stock" and that "the relations of the different groups should be capable of representation by a genealogical tree, or *phylum* as Haeckel calls it in his remarkable 'Generelle Morphologie'." Huxley drew up just such a tree, deriving all carinates (flyers) from a partridge–like ancestor, and he began reclassifying birds into "logical groups", telling the editor of *The Ibis* that this was "simply a first and most important stage in the progress towards the ultimate goal, which is a *genetic classification*", defining this as a classification "which shall express the manner in which all living beings have been evolved one from the other." All of which is strikingly reminiscent of Haeckel's goal – to "introduce the Descent Theory into the systematic classification of animals and plants, and to found a 'natural system' on the basis of genealogy . . ."[23]

1868, the year of *Bathybius Haeckelii*, probably marked the height of Haeckel's influence. Look again at Huxley's letter, written that January, in which he admits to be reworking the descent of dinosaurs. He tells Haeckel:

> The road from Reptiles to Birds is by way of Dinosauria to the Ratitae. The bird "phylum" was struthious, and wings grew out of rudimentary forelimbs.
>
> You see that among other things I have been reading Ernst Haeckel's *Morphologie*.[24]

Huxley's Social Strategem

> Both Darwin and Wallace – indeed, Herbert Spencer and Huxley as well – were members of the middle class before they were scientists. By the mid-nineteenth century this class had developed a cultural paradigm which assumed that progress was dependent upon economic competition. In the fifties cosmic optimism was supported by a rapidly rising gross national product.
>
> > Michael Helfand (1977) in an
> > essay on the political import
> > of Huxley's "Evolution and Ethics".[25]

We must dig still deeper to discover why Huxley found the *Morphologie* so attractive; after all, its leading tenets clashed with almost everything he had once taught regarding life's persistence. His use of a bird pedigree as a snub to Owen can hardly be the whole explanation of his adoption of Haeckelian science. Owen never invoked descent, resting his case on the morphological similarity of mammals and birds. So Huxley could simply have compared birds to reptiles without mentioning evolution and been just as effective, perhaps more so. Haeckel to many was the devil incarnate; putting his programme into effect might ultimately have cost Huxley votes.

This being so, we should investigate the growing socio-political use Huxley found for genealogy. But to do this we must first consider the historical setting, concentrating on the social inequalities and political solutions of the period (an unsettled one at that). 1866–1870 was a time of growing trade-unionism and demands for reform, suffrage and educational opportunities. Paradis has noted that the social theme in Huxley's essays strengthened during these years. So did his materialism, which is understandable since it was part of the anti-aristocratic, anti-clerical ideology of his 'mercantile' class. He actually portrayed the 'scientist' in his Working Men's lectures as a proletarian (which would not have endeared him to the Cambridge dons). The artisan and anatomist were labouring brothers, each having to grapple with often sordid reality at close quarters. In 1870 Huxley's essays were gathered under the title of *Lay Sermons*, and they make instructive reading. "On the Advisableness of Improving Natural Knowledge" had been delivered at St Martin's Hall in 1866 and centred on the technological benefits brought by science. Here Huxley argued that as the "laws of comfort" (i.e. technological improvement) were better understood, so too were those of "conduct". This theme was to be reiterated time and again – that science was the essential accompaniment of right morality and civil order. In "Emancipation – Black and White" (1865) he pleaded that negroes and women be freed for the better functioning of society, reassuring his audience: let the inferior compete in an open society, "Nature's old salique law will not be repealed, and no change of dynasty will be effected."[26] The message in "A Liberal Education: And Where to Find it" (1868) was equally clear – continued reform for the stability of society. The *Lay Sermons* thus served a variety of functions, but one wonders how many political masters. There is no doubting Huxley's sincerity in championing the workers' cause, but it seems he often had more than one end in view. His heartfelt sympathy for the poorer artisans is apparent in, say, his address to the Midland Institute on "Administrative Nihilism" (1871). Here he applauded the Education Act of 1870 and welcomed state intervention in this area, taking Spencer to task for his pernicious individualism, according to which not a penny from state coffers was to be spent on education, let alone poor relief or medical aid. Huxley insisted that it was crucial for social and economic stability to educate the poorer wage-earners and help the more gifted become socially mobile (conversely, he hoped to see the popping of corks that kept afloat the idle rich). In all these essays he pleaded passionately for political and educational reform, canvassing from a proletarian base. Indeed, he declared himself to a university audience in 1874 as "a plebeian who stands by his order". But however straightforward his social commitment seemed, I shall argue that it was inestimably more complex.

A related point is made by Paradis. It is that Huxley consciously broke with the Carlylean ideal of the hero in this period. The insidiousness of Carlyle's social ideals became manifest during the Jamaica affair in 1866. Governor Eyre in the colony had ruthlessly quashed a revolt by blacks and hanged their leader without trial. Huxley, Darwin, Lyell, and Spencer all joined Mill's Jamaica Committee, determined to see Eyre prosecuted for murder. Interestingly, Tyndall joined Kingsley, Dickens and others in supporting Carlyle's defence. Huxley told Kingsley, a friend of Eyre's, that he would as soon migrate to Texas as see heroes above the law. In practical terms, hero worshipping was a dangerous step towards despotism, whereas his own ideal of science was "a democracy of knowledge and a social leveling force".[27] According to Paradis, Huxley had moved significantly towards "an egalitarian ideal".

I think this undeniably true, yet still only part of the picture. While discussing worker education, he most often had in mind scientific and technical education; in other words he was spreading the scientist's power base to the classroom, and was actually lobbying on behalf of the middle-class scientists. He recognised that with educated and articulate workers assuming positions of greater power, science itself would become a more formidable force, and its middle-class exponents better respected and valued. It also seems to me that Huxley's essays were designed to appeal to bosses, and to persuade them to initiate workers into scientific modes of thought *for the stabilisation of capitalist society*. In short, he was wooing both sides from a middle position.

To appreciate the value of Huxley's stratagem one has to understand the context of social unrest and the capitalists' fears he played upon. One sees his complex double ploy admirably in "A Liberal

Education". He was talking *to* working men, but much of what he was saying must have been immensely comforting to the entrepreneurs and capitalists who subscribed to *Macmillan's Magazine*. We have seen that he equated a scientific education with a grounding in morality: understanding the natural order taught men to value good conduct. Thus the scientist was a guarantor of social stability at a time of radical demands and widespread agitation.

> A workman has to bear hard labour, and perhaps privation, while he sees others rolling in wealth, and feeding their dogs with what would keep his children from starvation. Would it not be well to have helped that man to calm the natural promptings of discontent by showing him, in his youth, the necessary connexion of the moral law which prohibits stealing with the stability of society – by proving to him, once and for all, that it is better for his own people, better for himself, better for future generations, that he should starve than steal? If you have no foundation of knowledge, or habit of thought, what chance have you of persuading a hungry man that a capitalist is not a thief 'with a circumbendibus?'[28]

He was speaking in 1868, at the opening of the South London Working Men's College. Only a year before demands for suffrage had finally forced Disraeli to steer the second Reform Bill through the Commons, thus doubling the vote and putting another million on the register (to the inexpressible disgust of Tory diehards). Unionism was rampant, and 1868 saw the establishment of an umbrella organisation, the Trades Union Congress. In France renewed republican agitation had led to a wave of strikes, forcing a number of liberal concessions (noticeably the lifting of press censorship and bans on public meetings – which only served to fan the flames and create the conditions for the Commune). Capitalists on this side of the Channel were naturally apprehensive, and by appealing for the "democratisation" of knowledge, Huxley hoped to convince them that investment in this sector was their best guarantee of political and social stability, without which future progress was impossible. There was nothing disingenuous in this. He was not selling one class short or treacherously appeasing another, but taking the only line consistent with his middle-class liberal principles. He was not agitating for revolution, but *reform* in which the conditions of the workers would ameliorate as his own social group, the bourgeois scientists, advanced.

Looking at the problem from the standpoint of Huxley's own political fears tends to reinforce the thesis. He saw science as a means of social levelling without socialist disruption. Paradis quotes a passage from the "Progress of Science" to illustrate Huxley's "egalitarian ideal":

> All these gifts of science are aids in the process of levelling up; of removing the ignorant and baneful prejudice of nation against nation, province against province, and class against class; of assuring that social order which is the foundation of progress, which has redeemed Europe from barbarism.

Actually Huxley does not stop there, but in the original continues:

> and against which one is glad to think that those who, in our time, are employing themselves in fanning the embers of ancient wrong, in setting class against class, and in trying to tear asunder the existing bonds of unity, are undertaking a futile exercise.[29]

Bearing in mind the date of this essay (1887), and the fact that the Fabians were already advocating State Socialism and Wallace land nationalisation, one begins to read the passage somewhat differently. Huxley was presenting science as an alternative to more drastic socialist remedies. Technology within the capitalist structure *could* satisfy the workers' moral and material demands; science *could* ensure financial security for wage-earners without the need of a class revolution. Indeed the "stable equilibrium of the forces of society" was itself the "foundation of progress', so a revolution might be counterproductive. Michael Helfand reminds us that we should understand the social setting if we really wish to appreciate Huxley's intent; and he points the way by reinterpreting Huxley's puzzling Romanes Lecture "Evolution and Ethics", setting it against the increasing attacks on Darwin and Malthusian economics (on which natural selection was based), attacks reflecting the growing Socialist movement in British politics. Helfand calls Huxley's essay a "masterpiece of concealed debate" – an attempt to vindicate the Darwinian status quo and "justify a Liberal imperial political policy" then under siege. But Huxley's opposition to socialism reached back to 1870 or earlier. In "Administrative Nihilism" he not only denounced extreme *laissez faire*, but conceded that lack of state education fuelled the flames of socialist discontent and could have dire consequences if the newly-enfranchised masses listened to their demagogues. "What gives force to the socialist movement which is now stirring European society to its depths," he asked, "but a determination on the part of the naturally able men among the proletariat, to put an end, somehow or other, to the misery and degradation in which a large portion of their fellows are steeped?"[30]

Since Huxley's recipe for social stability rested on the ability of educated workers to rise above their station, we can understand the political use he found for Darwinism and the Haeckelian interpretation of "phylogeny" or racial ascent. Huxley himself was deeply conscious of the bourgeois merit of being self-made, even if he joked to his wife in 1876, after an American reporter for *The World* had been delighted to find him "of the commercial or mercantile type", and not so "highfalutin" as he had feared: "This is something I did not know, and I am rather proud of it. We may be rich yet." The truth is he *was* doing extremely well for himself. Despite the financial handicap of being a professional scientist, he was earning over £950 p.a. by 1864, £2000 by 1871, and £3000 by 1876, about half of which came from his popular lectures. He did not have to look far for his image of the self-made man. Indeed, it may have been this self-betterment image which predisposed him to accept Darwinian evolution, stamping it in turn with his middle-class values and so making it readily accessible to a mass audience. Consider the way he chose to present the "burning" issue of man's origin, and especially the distasteful subject of his "pithecoid pedigree". He wrote in *Man's Place*:

> thoughtful men, once escaped from the blinding influences of traditional prejudice, will find in the lowly stock whence Man has sprung, the best evidence of the splendour of his capabilities; and will discern in his long progress through the Past, a reasonable ground of faith in his attainment of a nobler Future.[31]

Consider, too, that he was speaking to working men, "my"

working men, as he called them; the sort of working men who were attending Mechanics' Institutes, i.e. no longer solely the working-class elite, mechanics, artisans, and the like – by now many of these had abandoned the Institutes or been ousted by the middle-class clerks, shopmen, and book-keepers. And one realises that he was evoking a class image: charging them, in effect, to use the analogy of their own enterprise and ability to better themselves in order to grasp the meaning of evolution. Fast-rising middle-class professionals in the wake of Industrialisation would have understood better than anyone else what he was talking about. This explains why Huxley's books, like Darwin's, were almost 'mass market', and were even snapped up at railway stations, which caused consternation in more traditional circles. Certainly the 'lowly-ancestor, noble-future' image would have fallen remarkably flat among the landed gentry, Cardwell's "fox-hunting, port-swilling squirearchy". The geologist Robert Godwin-Austen acknowledged his complimentary copy of Huxley's book with the reassuring analogy: "Man's descent reads very much like a Law Lords pedigree in the Peerage – a remote ancestor temp. Will. I [at the time of William the Conqueror]. Seventy fourth in descent a Wig-maker – and then the full-blown Chancellor or Chief Justice."[32]

Godwin-Austen spectacularly caught Huxley's drift. He had tailored his presentation of evolution to meet specific social needs: the image of a humble wig-maker rising to Lord Chief Justice would have been immensely appealing in the 1860s. To the worker, evolution bestowed scientific dignity on his lowly parentage and promised salvation in this life. Obviously by the late 1860s 'persistence' was becoming hopelessly inadequate to Huxley's social needs – it did nothing to reassure workmen of their noble destiny, nor did it chime with bourgeois faith in social progress. Haeckel's exercise in "phylogeny" or racial progress promised a better return, and Huxley began to rationalise the changeover in his lectures. Although 'persistence' was to remain a prominent part of his geological thinking, the switch to "geologic" progression had clearly been initiated by the Reform year of 1867.

Back to Fossils – The Continental Evidence

Thus the Geological Society address of 1870 saw Huxley "softening the somewhat Brutus-like severity" with which he had formerly treated "progessive modification".[33] And he started to survey the evidence for the genealogy of vertebrate families; in fact, so much ground-work in reconstituting fossil lineages had been done – mostly by Continental non-Darwinians – that he could afford the luxury of picking from among the best.

There was no Darwinian Revolution in France,[34] but a number of palaeontologists had taken a particularly Gallic approach to *transformisme* in the 1860s, and come up with some convincing evidence. Albert Gaudry (1827–1908) from the Jardin des Plantes had led two expeditions to Pikermi (near Athens), in 1855–6 and 1860, and recovered the best part of a Miocene mammalian fauna, comprising 51 species and 374 individuals, many of which looked better suited to an ancient African plain. Among the primitive monkeys, cats, hyaenas, mastodons, rhinos, etc. (and less familiar chalicotheres and *Dinotheriums*) were a number of linking forms, e.g. the carnivore *Simocyon*, with its dog-cat-bear features; and *Mesopithecus*, seemingly midway between a macaque and langur monkey. The Pikermi fauna neatly filled the hiatus between Cuvier's archaic Eocene fauna of Montmartre and living types, and Gaudry in his multi-part *Fossil Animals and Geology of Attica* (1862–7) treated many of his fossil families – hyaenas, pigs, elephants, and horses – to a lavish family tree, drawn up from European, American, and Indian fossils. However provisional, the detail was impressive: the horse dynasty, for example, with *Hipparion* its Miocene head, could be traced into seven branches by the time of *Equus*.

The book was published to great acclaim. Huxley heralded it as "one of the most perfect pieces of palaeontological work I have seen for a long time." Darwin endorsed it sight unseen, telling the author he fully expected to find it "a perfect mine of wealth". Lyell had actually glimpsed Gaudry's fossils in Paris and was so overwhelmed that he at last felt able "to appreciate the force of the evidence appealed to in favour of Transmutation".[35] Extracts from the work quoted by Lyell in his 10th (and evolutionary) edition of *Principles of Geology* appeared to Darwin "the most striking which I have ever read on the affiliation of species." Gaudry only cautiously conceded that the palaeontological scales had tipped in favour of transformism, declining to speculate on the mechanism or even to discuss Darwin's haphazard variation. Nonetheless, Lyell admired the man's courage, acknowledging a kindred spirit – his own years of soul-searching, and particularly his struggle to rationalise a dryopithecine ancestry for man, had left him peculiarly sensitive to the ordeals of others.

Like Gaudry, the Swiss palaeontologist Ludwig Rütimeyer (1825 –1895) had issued a series of monographs, methodically executed and mixing a vast amount of meticulous and detailed description and deductions on the probable lines of descent, especially of wild and domestic oxen. He occasionally tackled controversial matters, for instance, in a little paper in 1868 on anthropoid apes with the self-explanatory title, "The Limit of the Animal World: A Consideration of Darwin's Doctrine".[36] But generally he steered clear of Darwinism and its implications and most of his work at Basel (where he was Professor of Zoology) concerned the origin and racial variation of domestic livestock.

Thus by the time Huxley had warmed to "progressive modification" he had a choice of fossil lineages to popularise. He was noticeably kinder to younger Continentals than older non-Darwinians at home, partly, I suspect, because they refrained from passing adverse judgment on Darwin. He called Gaudry and Rütimeyer "two learned, acute, and philosophical biologists".[37] Since, from their work, it was apparent that the horse pedigree would withstand "rigorous criticism", he spent some time detailing the route by which today's one-toed *Equus* was derived from the small Middle Eocene three-toed *Plagiolophus*. And in the process he indulged in a celebrated piece of "retrospective prophecy", as he later called it, arguing that the two outer splints were remnants of once-functioning toes and that therefore a five-toed ancestor would sooner or later turn up.

None of this was very new. Even Owen had used horses to exemplify his theory of progressive "departure from the general type". As early as 1851 he had constructed a series from the old *Palaeotherium* of Montmartre to the Pliocene *Equus*, explaining that the horse had grown "swifter by reason of the reduction of its toes"[38] until it could run on a single digit. But Huxley's relationship with the Grand Old Man was so delicate that he would never have admitted putting an evolutionary gloss on Owen's sequence. We have seen that in 1860 Huxley had actually denied that *Palaeotherium* was more "generalised" (trying not to play into Owen's archetypal hands), even if, he said, it was the "father" of later forms. Easing into his new Darwinian role, Huxley still avoided such loaded terms as "generalised", but now admitted that the route leading to the modern horse "is one of specialisation, or of more and more complete deviation from what might be called the average form of an ungulate animal". For "average" read "general", there was no difference. Owen for his part continued to argue for the "derivation of equines", for example in his *Anatomy of Vertebrates* (1868). And here he actually used an illustration of the sequence of teeth and toes that could have been a forerunner of Huxley's famous and much-copied diagram in *American Addresses*, showing once again that, despite the ideological switch, there was structural continuity between pre- and post-Darwinian palaeontology.

Huxley paraded the horse as "The Demonstrative Evidence of Evolution" in *American Addresses*, by which time he had incorporated Marsh's New World fossils. But it could be argued that he had actually done less than Owen to explain why post-Eocene horses had grown steadily taller and more thoroughbred-looking. Functional explanation was never his forte, perhaps because of his notorious disrespect for Cuvier, or the fact that he lacked an education in early-Victorian natural theology which looked to designed adaptations as proof of Beneficence and Foresight. These influences explain the crucial role accorded specialisation and adaptation in the *Origin*, but as Rudwick observes, the ecological emphasis faded rapidly after 1860 with the rash of lineage tracing:

> Most palaeontologists forgot – at least in practice – that their fossils had once been living organisms, and that they had been adapted to some mode of life; and they tended to regard their specimens exclusively as evidence for evolutionary ancestry.[39]

Not everyone forgot. While palaeontologists eagerly sought progressive lineages, a physiologist from University College, Huxley's young friend Michael Foster, reminded *Quarterly* readers in 1869 that:

> The further we extend our survey over the Invertebrate Kingdom, the oftener is the idea of an arbitrary progressive

impulse, driving the animal form from lower through higher phases towards a standard of dignity, obliged to give way to the conception of a benign influence, moulding the plastic organism to fit into every variety of circumstance, and to make the best of all the accidents of life.

This kind of deconsecrated Paleyism was at the very heart of the *Origin*. Sedgwick's caring Father, personally and immediately responsible for the fit of every mollusc or fish into its locally-available niche, had become a "benign influence" (natural selection) which made the most "of all the accidents of life". But later Victorian palaeontologists evidently found this an impractical approach, and the Darwinian emphasis risked being lost in the scramble for lineages and pedigrees.

Haeckel had coined the word "ecology" to replace the old 'economy of nature', although he did little if anything to further its understanding. But he (or Gegenbaur) attracted to Jena a young Russian dissident, Vladimir Kovalevskii (1842–1883), who was finally to give adaptation its due. A number of scholars have already tackled Kovalevskii's innovative approach to Darwinian palaeontology from the perspective of his revolutionary background (he was greatly influenced by the exiled revolutionary Alexander Herzen). The shock of defeat in the Crimea (1853–6) had caused a political reawakening in Russia and widespread demands for social reform; but Tsarist measures, like the emancipation of the serfs in 1861, had actually done little to ease social tensions, and the militant intelligentsia embraced Darwin for demonstrably subversive ends: seeing the lawful and materialist implications of the *Origin* mock Orthodox spiritualism and damn the "irrationalist bent of autocratic ideology".[40] Kovalevskii himself was a widely-travelled publisher of radical books. He eagerly exploited Western materialist science to sanction revolutionary change and, like fellow activists, appreciated the political importance of Darwin's 'adapt-or-die' thesis. He acknowledged the "complete revolution caused by DARWIN's great work", visited Down twice during his travels (1867 and 1870), and helped translate Darwin's *Variations of Animals and Plants* and *Descent of Man* into Russian. And when he turned his own hand to palaeontology, it was this 'adapt-or-die' thesis that made it so convincingly different from any other nationalistic form.

Both Gaudry and Huxley accepted that evolutionary mechanics

were the province of the physiologist, not the palaeontologist. For Kovalevskii, however, adaptation and competition were an integral part of palaeontological explanation; he found it imperative to re-associate life and environment to explain why certain fossil routes were followed. For instance, he saw the hoofed mammals develop along competing lineages, with victory going to the ones best adapted to environmental conditions. He illustrated this with an elegant study of "adaptive" and "inadaptive" modes of toe-reduction in the lines leading to modern ruminants, correlating the various permutation of joints with survival potential, and plotting the eventual demise of the inadaptive types. Likewise he dealt with the horses, relating crown size in the molars and gradual loss of the toes to the evolution of grassy plains in the Miocene. By refocusing the question, Kovalevskii produced the most innovative monographs on ungulate evolution in the late nineteenth century – and this in an astonishing short space of four years (1869–1873) – after which he abandoned palaeontology, struggled to find a teaching post, and finally chloroformed himself to death.

So there was a wealth of material from diverse traditions awaiting reorganisation from an evolutionary perspective, and most English Darwinians, when they were not being wholly descriptive, preferred synoptic to innovative studies (Huxley's dinosaur-bird ancestry being an exception). Flower is a case in point. His Friday evening lecture to the Royal Institution "On Palaeontological Evidence of Gradual Modification of Animal Forms" (1873) followed Kovalevskii on artiodactyl toe-reduction, placed Leidy's American oreodonts (the "ruminating swine", totally extinct) in an evolutionary context, and looked at rhinoceros specialisation, but in no greater detail than had Owen twenty years earlier. He endorsed Gaudry on horse evolution, and like Huxley declared this "more like a real genealogical history than any other known".[41] From these facts Flower concluded no more than that life was modifiable and that "the more modern specialized forms converge towards the ancient more generalized forms", which would not have come as much surprise to Owen.

The Evolutionary Tempo – Huxley on Crocodiles

Finally I want to turn to the 'crocodile episode' (1875–1879), to show the practical upshot of Huxley's evolutionism and Owen's alternative to "phylogeny". Here for the first time we hear of a varying rate of change within a single lineage. Why Huxley might have found a changing tempo acceptable is not hard to see; most obviously, it still denied the kind of inexorable ascent so often associated with transcendental explanations, and thus it was not incompatible with his earlier Lyellian stratagem. But it also meshed nicely with his conception of social history, which consisted of long stationary periods (invariably caused by a 'dogmatic' Christianity, e.g. the Dark Ages) punctuated by Renaissances, which he associated with a flowering of the rationalist spirit (such as the Victorian era, with its promised "new reformation"). Metaphors were liberally exchanged between history and geology, and even biological parallels were sought – for instance, he likened history to an insect undergoing periodic ecdyses, or moults, times of liberation from old constraints and renewed growth. So history, as interpreted by a mid-Victorian secularist, could have provided a fruitful analogy for punctuated fossil progression. Flower, while admitting "a general progress [of life] from lower to higher or more specialized types", cautioned that this was, "as in all other cases of progress (human civilization, for instance), attended with many exceptions, some local and temporary, some only apparent."[42]

On the crocodile question, Huxley began typically with an unquestioned allegiance to Lyell, valuing these primeval reptiles for their ageless appearance. Between 1859 and 1870 he asserted time and again that early Mesozoic crocodiles were "identical in the essential character of their organization with those now living."[43] In 1870 he did admit of "very small differences", but only to ponder the "appalling" span of time since the group's emergence (derived by extrapolation, assuming a *uniform* rate of change). But by 1875 he had changed tack entirely. Convinced of the capital to be got from a progressive crocodile sequence, he dropped all allusion to 'persistence' and presented the armoured reptiles as a test case for evolution. Now he argued that crocodiles had put on a Mesozoic spurt, progressing through a number of identifiable stages before sinking back into quiescence.

It was not so much new fossils as new theoretical concerns which caused the shift. He had been studying the same crocodiles (or what he took to be crocodiles) since 1859, when he identified the sixteen-foot *Stagonolepis* from Elgin as the earliest example. Even then he noticed certain aberrant features, but failed to modify his repetitious statements on the group's 'persistence'. Admittedly, the bones in his possession were friable and far from perfect; nonetheless, it was the Lyellism which dictated his approach. Throughout the sixties, new consignments of Elgin reptiles were shipped to the Museum of Practical Geology by Rev. George Gordon (1801–1893), the Presbyterian minister at Birnie in Scotland, who received a Royal Society grant for this purpose. Gordon was a grass-roots naturalist, a collector of rarities and antiquities, and according to the *Moray and Nairn Express*, at least, second only to Hugh Miller as the North's most gifted geological son. He meticulously searched the Elgin quarries and dispatched to London specimens of the "lizard" *Telerpeton*, the "crocodile" *Stagonolepis*, and an unknown Tuatara-like reptile, which Huxley named *Hyperodapedon Gordoni* in his honour.* From these fossils Huxley built up a pretty fair picture of *Stagonolepis* and by the time he published on crocodile evolution in 1875 was no longer prepared to gloss over differences between this and more modern forms. He proposed to group all living and fossil crocodiles in three suborders. The lowest – typified by *Stagonolepis* – he called the Parasuchia. Next in line stood the advanced Mesosuchia, with a developing secondary palate, and finally the more modern Eusuchia with a complete secondary palate (the bony roof of the mouth, separating the nasal cavity). He presented these results almost as an exercise in logic, reasoning that if the morphological series had really been wrought by evolution, these stages should appear sequentially in the rocks – which of course they did; the Eusuchians finally emerging in Cretaceous times, after which there was little or no advance.

Huxley's series was not as modern as it seems, for he actually

---

* There was an added incentive to hunt down these precious fossils. The Elgin quarries had long been considered Devonian, so the unexpected discovery of *reptiles* gave the beds an unprecedented importance, not only locally, but internationally. Reptiles were simply unknown from such antiquity. Gordon, perhaps as a patriotic gesture towards Morayshire, "the fairest spot on earth",[44] led a band of local geologists in defending a Devonian age, and went to his grave at 92 convinced that London geologists were wrong in raising it to Triassic. Ironically, the prime evidence for this reinterpretation were the reptiles – including *Stagonolepis* – that Gordon himself dispatched to Jermyn Street.

envisaged a progressive "departure from the Lacertilian" towards the crocodilian type. In other words, crocodiles, like dinosaurs, diverged from a 'lower' lizard-like base; and with consternation in some quarters over just what constituted "higher" and "lower", he argued analogically from contemporary technology that crocodiles "are higher than lizards as a steam-vessel is higher than a sailing-ship",[45] i.e. an advanced modification. So his overall picture depended both on the ascent of crocodiles and stability of lizards. He had already diagnosed another Elgin reptile, *Telerpeton*, as a ten-inch lizard and "one of the most astonishing examples within my knowledge of a *persistent type*". It apparently presented no Triassic features, nor was it "a 'generalized' form, or as, in any sense, a less perfectly organised creature than the Gecko". Huxley it seems never quite broke free of his Lyellian moorings, and was happy to place modern lizards in archaic settings, despite the protestations of young turks like Harry Seeley.

Otherwise the paper was acclaimed, and Martin Duncan actually thought it "furnished a stronger support to the hypothesis of evolution than even that of *Hipparion* and the Horse".[46] More revealing, however, is the long-term reaction, for it highlights a profound methodological split between the Darwinians (Huxley and Hulke) and senior palaeontologists like Owen. Huxley showed no sign of interest in factors which might have explained his sequence. Despite being instrumental in getting Kovalevskii's monograph published by the Royal Society, he seems to have remained indifferent to what Osborn called the new palaeontological "spirit of recognition of the struggle for existence, of adaptation and descent." John Evans did ask whether the migration of the internal nostril could "be correlated with differences in the mode of life?" Huxley had evidently considered the obvious answer – that the secondary palate separating off the nasal passage allowed crocodiles to drown their prey without shipping water; but he was inclined to dismiss it because it failed to apply to fish-eating gavials. The truth is, it wasn't his problem.

Hulke responded to Huxley's initiative by interposing yet another waystage (the Metamesosuchia) on the crocodile's path, inserting the Purbeck fossil *Goniopholis* to give it more the feel of a continuum. However, older practitioners steeped in Cuvierian functional anatomy and the natural theology of animal design adopted a quite different approach. Owen in his *Palaeontology*

(1860) still considered those who were blind to the beauty of design as suffering from "some, perhaps, congenital, defect of mind". Eleven years later in his *Monograph* on Mesozoic mammals he pitched into such "beginners in Palaeontology" (!) as Flower, Dawkins, and the late Hugh Falconer – then haughtily proclaimed himself "a disciple of CUVIER" (i.e. infallible), and warned of the dire consequence if the Cuvierian light were extinguished.[47] This was a gross miscalculation on his part, for *le baron* was fast becoming the whipping boy of Darwinian historiography. Huxley grudgingly conceded the man's genius, but blamed him for the "evil times" visited upon Lamarck; Haeckel proclaimed his views "the greatest obstacle" to a true history of creation, and Marsh frankly conceded that Cuvier's authority had "delayed the progress of Evolution for half a century".[48]

Yet the Cuvierian perspective had its advantages. Hulke introduced *Goniopholis* at the Geological Society on 6 February 1878. Owen, now a sprightly seventy-three, appeared at the same meeting, dumbfounded to learn that he and Hulke had been working, not merely on the same crocodile, but on the same specimen (which speaks volumes for his continuing alienation). Unlike Hulke, Owen preferred to fit *Goniopholis* into one of the pre-existing categories. But not out of cussedness, nor necessarily to thwart the evolutionist; being concerned with the balance of the ecosystem, he needed to break up the sequence into discrete units – to freeze the frames, as it were, in order to tease out the causal relations. What we would call the environmental dimension, Owen unfortunately called the "teleological" (something of a dirty word in post-*Origin* days), i.e. the relationship of structural modifications to "concomitant changes in external influences."[49] He concluded that the stiffly-spined, smaller-limbed crocodiles gave way to larger-limbed, more-manoeuvrable forms as the first big mammals browsing along the shorelines tempted "them to make a rush on the dry land". An entirely new source of prey had forced wide-ranging structural changes. Even the appearance of stabbing canines was correlated with the new diet, while the secondary palate allowed the creature to breathe while drowning its prey. And following jibes by Hulke, who observed that tiny mammals had *always* existed alongside crocodiles, Owen triumphantly produced newly-discovered dwarf crocodiles like *Nannosuchus*, some with skulls barely four inches long, small enough to tackle even shrew-sized "marsupials". Purbeck crocodiles had miniaturised in response to these "succulent morsels" at the riverside, leaving a perfect predator–prey balance, but on a reduced scale.

On the whole this kind of ecological determinism was poorly received. Owen's paper was criticised not for what it did do, but what it didn't – namely mention "evolution" (a fashionable word he studiously avoided, talking only of the "coming in of the mammalian class" as the crucial factor). Huxley's supporters were not concerned with functional suites, nor with new hunting habits which might explain them; their Haeckelian priority was clearly to establish an acceptable pedigree.

From this it would seem that providential science by now had its own goals, as a result of being constructed differently at both the conceptual and social levels. In fact, its contrasting social base suggests that it was designed to support the aspirations of a rival section of the community. Looking at it from this perspective might help explain why extreme transcendentalists side-stepped phylogeny altogether, avoiding the kind of genealogical approach favoured by Huxley and his circle. Providential palaeontology certainly deserves a more sympathetic study than it generally gets, and this we shall attempt now.

6

Groves of Trees & Grades of Life
The Idealist Legacy

Polyphyly and Progressive Evolution

Stereotyping the "creationists" in an effort to expunge metaphysics from science, Huxley helped create the resilient image of what the first president of Cornell University, Andrew White, evocatively called *The Warfare of Science with Theology*. This powerful "military metaphor" and the Whig approach to history fostered a blatantly partisan appreciation of the past. The tendency was, and sometimes still is, to dismiss the post-Darwinian idealists as obscurantists and reactionaries, the villains of the piece, their only function to harry and hinder, or muddy the philosophic waters by resorting to "pseudo-scientific realism".[1] Here I want to take a more positive view. I want to continue the theme that Platonists, however unsuccessful in the alien realm of evolutionary mechanics, did contribute to the conceptual foundation of palaeontology – their belief in purposeful variation and polyphyly (descent from multiple ancestral stocks) stemming directly from an ideal Christian worldview dominated by the Divine Thought. Further, I shall suggest that apparently pure or value-free discoveries – particularly of a reptilian ancestry for mammals – made within this ideological climate were tied to it for their ultimate meaning. In this way I shall attempt to explain why a protagonist like Huxley, so utterly and fundamentally at loggerheads with any form of transcendentalism, could find such pre-empted reptilian ancestors difficult or impossible to accommodate, and as a result be left searching for an alternative pathway involving more compatible forms.

Mivart's desertion from the Darwinian camp was symptomatic of the general disillusionment among idealists with the *Origin of Species*. Like many others, he could never accept that random or non-purposeful variations formed the material for environmental selection. The new science of design reflected a growing desire for order in the face of this "law of the higgledy-piggledy", and because it promised to replace Darwin's fortuitous base with a lawful directional change of species, it was hailed as "*the* rational view of the universe".[2] The kind of teleology which saw life preordained and developed according to plan was ultimately dependent on a Christian conception of causation, in which efficient causes were equated with active agents. Indeed for Chambers and the Duke of Argyll, laws – no less than causes – were real entities, and the instruments of Divine legislation. The Word of God was thus fulfilled and nature constantly reformed by *The Reign of Law* (the title of Argyll's best-selling book), with all its implications of Judicial rule. The Duke himself was a leading "legalist of nature" and actually wrote the book while trying to juggle Gladstone's Reform Bill through the Commons in 1866.[3] For Mivart, an "innate drive" triggered by changing conditions generated new variations, on the analogy with "the way crystals (and, perhaps from recent researches, the lowest forms of life) build themselves up according to the internal laws" of molecular aggregation. Because designers resorted to this kind of crystal or inorganic analogy and stuck to the letter of the law, while branding Darwin's model chaotic and dangerously Lucretian, they also prided themselves on being the better scientists.

Roger Smith rightly points out that to make sense of Darwin's critics we need some "definable notion of scientific community".[4] Whether Owen, Mivart, and Argyll formed the nub of one such sub-culture is a moot point. This group was never as homogenous as, say, the 'Cambridge Network' of the 1830s, comprising Whewell, Herschel, and Babbage, or the 'scientific naturalists' of the 1860s. But even the *x*-Clubbers, for example Huxley and

Tyndall, notwithstanding their common professionalising aims, held such "fundamentally different political principles" as to place them on opposite sides during the Eyre controversy and the American Civil War. And taking Spencer's *laissez faire* convictions into account, the 'economic' range of the group was indeed vast. But something similar might be said of the idealists. They shared a set of common goals, despite religious, educational, and class differences. Actually these are not as great as first appear. Although the Duke of Argyll was a Scottish Presbyterian, his Calvinism was "so eroded by new intellectual currents . . . as to have no bearing on [his] conception of evolution". Mivart was an old Harrovian and an influential lay Catholic, liberal to the extent that he was eventually excommunicated and his 'Mivartian' theory purged by hard-liners. And Owen was a Broad Churchman who, despite only a Grammar School education, promoted himself up through the ranks. So there is little doubt where the group's support came from. As the materialists appealed to the increasingly-powerful bourgeoisie – the main benefactors of the industrial boom – so the Owenians, like their Cambridge contemporaries, were largely backed by clerical, landed, and upper-class interests, in fact by those destined to lose most from industrial expansion and the accompanying secular movement. Perhaps we can even talk of a limited 'tradition', insofar as Mivart and Argyll learned their archetypal science in Owen's lectures at the College of Surgeons during the 1840s and 1850s. The London idealist community might have been less effective for the lack of anything like the professionaliser's power base, but it was nonetheless recognisable from the mutually-reinforcing books, reviews and correspondence (much of which remains to be published).

If the idealists shared a common problem, it was by no means restricted to their circle. Spotlighting the real cause of origin of the "favourable" variants was the prerogative of a much wider providential movement. The "theologian's theologian" J. B. Mozley insisted in his *Quarterly* "Argument of Design" (1869) that natural selection cannot create the variations it chooses between; these must spring from some "unknown reservoir", which is the real "*productive* agency". Argyll invoked a hitherto unknown "law" to keep "those variations in a definite direction", envisaging evolution on the lines of embryological development with its purposeful aspect. "The physical means by which that purpose is secured remain as dark as ever", he confessed in *Reign of Law*, though it was evidently deterministic, and "like all other physical forces, working to order, subject to direction, and having that direction determined by foresight, forethought, and contrivance."[5] By 1868 when Argyll was appointed Indian Secretary, the book had reached its fifth edition, although Darwin remained unimpressed: "clever", he had to concede, "but not very profound". Huxley did not even bother to read it on publication; only in 1887 after being challenged by Argyll to defend the Darwinian "Reign of Terror" was he forced to plough through it, telling the editor of *The Nineteenth Century* that he had never before been so aware of His Grace's ignorance.[6] Being a 'critical positivist', Huxley of course settled on the Duke's lamentable philosophy, pointing out once again that laws *cause* nothing, neither the fall of apples nor the descent of men.

But the idea of an active agent, a Will, as it were, reflected in causation, was immensely appealing to non-Darwinian Christians. Not unexpectedly, Owen hailed *The Reign of Law* as "a timely & wholesome antidote to the mischievous fallacies" of the positivists, and Argyll became "the qualified Metaphysician" to rebuke the "Positive" assumption that "all phenomena, without exception, are governed by invariable Laws with which no volitions, natural or supernatural, interfere."[7] Owen penned an appreciative fifty-two page review in 1867 (destined for the Tory *Quarterly*, but apparently never published) which applauded the Duke's vindication of type and teleology as much as his "intelligible definition of . . . the term 'Law'." Peter Bowler and the embryologist Jane Oppenheimer suggest that Owen's archetypalism was never strong and by this time almost extinguished,[8] but he went to great pains in this piece to censure "advanced Positivists" for treating the "ideal Archetype . . . with great dislike, not to say disgust", and he rounded off with a spirited and vigorous defence of what he still saw as one of the majestic conceptions of modern science. We can guess why he chose this moment to answer back. From his point of view, the British Association meeting in Nottingham the previous year had been a disaster, with the President William Grove scornfully dismissing "the very notion of the Type". And according to Owen such views "were in fashion at that meeting". But he had other grievances against positivists, besides their "contemptuous relegation of abstract theories, archetypal conceptions, final causes, & the like, to the 'metaphysical', or worse, the 'theological' phase of

Human thought". He found their aggressiveness appalling and accused certain "young Professors" of gaining a name for themselves by adopting "extremist views" rather than earning their reputation through hard work. It also rankled that such ungodly philosophy should be peddled in "popular lectures addressed to youth of both sexes, and to classes of Working Men at St Martin's Hall and elsewhere" (in the same vein Owen had berated Huxley in 1860 for hoodwinking wage-earners gathered at the Royal Institution into believing they were the sons of gorillas). The cry "indoctrination" had obvious political overtones and was shortly to culminate in the accusations of social abandon levelled at Darwinians during the Paris commune scare of 1871.

Most Christians were committed to directional variation for reasons more aesthetic than necessary. Omnipotence could have programmed an interminable succession of 'random' variations and let nature do the selecting, but it seemed wretchedly uneconomic and clumsy. Purposive evolution also speeded up the process. It could be fitted into the short geological timespan of 96 million years, which Phillips had calculated from sedimentary deposition rates (the physicist Sir William Thomson made it 100 million since crustal condensation). This amount of time was "utterly inadequate" for Darwin's hit-or-miss method according to Thomson's colleague Fleeming Jenkin, the Professor of Engineering at Edinburgh. To make matters worse, another Scottish collaborator, P. G. Tait, actually thought Thomson had been too generous and dropped the figure in 1869 to an appalling "*ten* or *fifteen* millions".[9] Had it been taken seriously, this would have thwarted almost any evolutionary accommodation (which is probably what it was designed to do).

So there was widespread acceptance of programmed change in the sixties and seventies because it permitted an ever-active Deity in a world of lawful uniformity. Even Charles Lyell agreed with Argyll that natural selection was a "force quite subordinate to that variety-making or creative power to which all the wonders of the organic world must be referred." This was one of the points the writer Samuel Butler made time and time again in his popular books – but then Butler, like Mivart, was made acutely aware of the materialist exploitation of Darwin. "It is not the bishops and archbishops I am afraid of", he once said. "Men like Huxley and Tyndall are my natural enemies . . ." In *Evolution Old and New*

(1879) he reduced the *Origin* to little more than "a piece of intellectual sleight-of-hand", professing to show us the way but really blindfolding us at each turn. What survives Darwinian selection "was born fit", and what caused it to be born fit was still theology's question. "Have you read Butler's '*Evolution Old & New*'?" Mivart asked Owen. "There is method in his madness & it will I think help to burst the inflated bubble of 'Natural Selection'."[10]

So the 1860s and 1870s saw a number of leading scientists and writers rebel against "blind variation", forcing Darwin to carry out cosmetic surgery on successive editions of the *Origin*. He particularly suffered from the mutually-reinforcing positions of Mivart and Jenkin. Jenkin's cruelly logical *North British Review* article (1867) silenced even a usually noisy Huxley – who admitted its "real and permanent value", and then promptly dropped the subject. Jenkin argued that any slight variation, however "favourable", would be blended-out in an ocean of normal peers, apparently convincing Darwin that solitary variations could not survive. Only if no blending occurred, if the "sport" bred true, could selection begin to work on it. But that, said Jenkin, was not Darwin's position, nor was it far short of a "theory of successive creations".[11]

Mivart compounded the problem in *Genesis of Species* where, according to Wallace, the arguments against natural selection as the sole cause were "exceedingly strong".[12] Mivart's letters reveal a man in awe of Darwin. He genuinely loved and respected the old naturalist, and once confessed that "it has been & ever will be, a most painful effort on my part to force myself to state my dissent from one who has so many just titles to my esteem". Yet that is what he felt compelled to do, and their relationship deteriorated in the most desperately unhappy circumstances, with Darwin growing angrier and Mivart denying that he was "biased by an *odium theologicum*". There was no overcoming their grave disagreement, and as Mivart told Owen in 1871, "The more I think over natural philosophy the more I feel that it is *necessary* to ascribe to bodies the possession of some *internal* principle regulating their evolution . . ." For Mivart life pushed forward in a well-planned drive. He denied Darwin's premise that transmutation was gradual, working on minute changes, pointing out that no retinue of fossil beasts, imperceptibly shading one into another, had ever been found. Nor was it probable that a half-way house, say an incipient wing, would be functional. Like Huxley and Owen he opted for a saltatory mechanism, but saw the modified organs make their appearance ready balanced and synchronised, something which had to happen in all individuals simultaneously. Only a guiding law working towards "harmonious, self-consistent wholes" could achieve effects of such magnitude across an entire wave-front,[13] and he took it for granted that the law itself was a consequence of "that innate potentiality which God has implanted" in living matter.

This much is well known, but on turning to working geologists we witness support coming from the unlikeliest places. Even a number of Huxley's colleagues and those he helped into professorships were sympathetic, although the fact is not immediately apparent. Few geologists made such a song-and-dance of their dissention as Mivart.

Take the case of William Boyd Dawkins (1837–1929), whose speciality was cave deposits and Pleistocene mammals. From 1861–9 Dawkins worked at the Geological Survey, mapping the Wealden formations of Kent and the Thames Valley, while learning curatorial techniques in the Museum of Practical Geology. Here he rubbed shoulders with Huxley, on whose recommendation in 1869 Dawkins was made curator of the Manchester Museum, and in 1874 he took the Chair of Geology at Owen's College (where Huxley was Governor). Knowing nothing else, we might easily misinterpret Dawkins's early statements. For example, he spent the 1860s disentangling Pleistocene rhinoceroses, and in the *Natural History Review* for 1865 admitted that the curious relationship of one- and two-horned rhinos was "incapable of any other solution than that offered by Mr Charles Darwin's 'Theory of descent with modification'."[14] Perhaps this was a sop to Huxley who, like a true editor, can be imagined asking how his results were to be explained. But given that Dawkins was the son of a Welsh clergyman, studied Classics at Jesus College, Oxford, and was converted to geology by John Phillips, himself doubtful of the effect of transmutation above the species level, one might have anticipated *some* reservations. Just how deep they ran is only apparent on tracking down anonymous reviews. Dawkins, it transpires, was the *Edinburgh* reviewer of both Darwin's *Variations of Animals and Plants under Domestication* (1868) and *Descent of Man* (1871). Here he supports evolution, offering Gaudry's fossil intermediates from Greece and Huxley's bird-like dinosaurs as evidence, but of the "trinity of causes" – selection, variation, and heredity – he holds up variation as the real creative element and points out cases, in the production of peacocks and pheasants, where selection can have played no part at all. Moreover he tentatively accepted Spencer's Lamarckian explanation – that environmental fluctuations *cause* functional changes within the organism, themselves "the primary and ever-acting cause of that change of structure which constitutes variation; and that the variation which appears to be "spontaneous' is 'derivative and secondary'."[15]

On *The Descent of Man* Dawkins was considerably less charitable. He denounced Darwin's "logical power" and thought that "most earnest-minded men", on learning that morals were the better part of brute instinct, would "be compelled to give up those motives by which they have attempted to live noble and virtuous lives".[16] With Paris aflame in 1871, Dawkins was not unduly paranoid in seeing a social threat and destabilising effect in Darwin's *Descent*. *The Times* also deplored its "disintegrating speculations" and declared the book "more than unscientific – it is reckless", coming at a time when "every artificial principle of authority seems undermined." English providentialists were understandably sensitive during the siege, although it affected some more directly than others. As a shareholder in the *Société Anonyme* or joint-stock company controlling the Paris *Jardin d'Acclimatation* – an institution modelled on London Zoo – Owen saw his investment wiped out as the exotic quadrupeds, even the prized elephants, were systematically slaughtered to feed the starving citizens. (Or rather the rich; evidently few of the *communards* tasted the morsels.) With aristocratic confidence, Mivart decried the revolution as a product of "loose" or atheistic philosophy. And he took the opportunity of drawing Darwin's attention to the "unnecessary irreligious deductions" being drawn from natural selection, ending with the apocalyptic cry: "God grant that we in England may not be approaching a religious decay at all similar to that of the middle of the 18th century in France which Frenchmen are now paying for in blood & tears!"

Dawkins' review in the *Edinburgh* was the perfect complement of Mivart's in the *Quarterly*. Both denied that selection was responsible for our higher faculties, or even that "man [as a physical entity] has been evolved from the higher apes through natural selection, although he were genetically descended from them",[17] which raises the inevitable question of why Huxley chose to "pin out" one rather than the other (see below). Given Mivart's example and pushed perhaps by political events, Dawkins toned down his Spencerian explanation and asserted that evolution was no "series of accidents". It presented a "nobler view of the great Creator" precisely because it was impossible "without the will of a directing Intelligence". At times this could almost have been Mivart talking: "without definite purpose", wrote Dawkins, "it is hard to believe how the simultaneous changes in one direction could be effected, and it is incredible that they should have been brought about by a combination of chances." All in all – and granted that he might have been shifting to the Right in response to the uprising across the Channel – it is still doubtful if Dawkins really supported Darwin's chaotic model in 1865. These later statements suggest that he was simply endorsing the broader idea of descent.

The repercussions for palaeontology of this internally-powered, 'wave-front' advance were spelt out by Mivart. With directional programming, it was possible for species from whatever source to proceed towards the same goal. Not only was the near-identity of many marsupials and placentals apparent proof of this, but Mivart thought that lemurs and apes also had separate origins, as did Old and New World monkeys. No longer was "similarity of structure" any guarantee of "genetic affinity".[18] Mivart had begun to dissociate "homology" from common ancestry as early as 1870, in reply to Lankester's anti-Platonic paper. To Mivart's mind, "homology" transcended Lankester's "homogeny" (similarity through shared ancestry) because it related *all* 'identical' forms, i.e. those having reached the same morphological goal, *irrespective* of their evolutionary origin. The same logic prevailed in Germany, where between 1872–5 the idealists Albert von Kölliker and Alexander Braun "attempted to save the idea of type and give it an autonomous position in biology, independent of the idea of descent." Kölliker reasoned that with physiochemical laws determining the route, branches from several genealogical trees might become intertwined, while Mivart argued in 1873 for a "grove of trees, closely approximated, greatly differing in age and size, with their branches interlaced in a most complex entanglement."

Mivart might have seen parallel development and convergence everywhere, but it was tactically important for him to do so. The branches of Haeckel's elaborate tree of life always met in a single stem; in other words, related groups were always derived from a single ancestral stock (*monophyly*). These gnarled family trees illustrated books which toed a 'dangerous' monist line; indeed, they became an inextricable part of Haeckel's whole argument. Mivart's rival *polyphyly* (in which a group could evolve from a number of unrelated ancestors) was intended to destroy the force of Haeckel's palaeontology and discredit his entire monist approach. Because "the number of similar structures which have arisen independently is prodigious", Mivart predicted that "many of the genealogical trees which have been developed with the rapidity of the fabled 'bean-stalk' are destined to enjoy an existence little less ephemeral."[19] As late as 1888 he was still making "sensational" claims for the multiple origins of the mammalia. He responded to Edward Poulton's unexpected discovery of true teeth in the young platypus by suggesting that it had a separate reptilian origin. The egg-laying monotremes were thus "hypothetical higher mammals in the making . . . more or less parallel to but, of course, radically distinct from, the placental and marsupial series", which, he assumed with Huxley, had an amphibian ancestry.

Poulton was a Darwinian and quickly scotched the idea. But Mivart's "grove of trees" was not unattractive to palaeontologists. P. Martin Duncan (1824–1891), for example, was a specialist on the affinities of ancient faunas, an area where one might expect a practical test of Mivart's ideas. Swiss educated and well provided for, Duncan received his medical training at King's College, London, after which he took up practice in Colchester, where he eventually became Mayor. However he threw over this lucrative practice to return to Blackheath in London to concentrate on the study of corals and echinoderms. In this field he became preeminent and his *British Fossil Corals* was said to have been one of the finest memoirs ever published by the Palaeontographical Society. His success can be judged from his taking the prestigious Chair of Geology at King's, where, according to the Council Minutes, he delivered his first six lectures on palaeontology in April 1869.[20] According to Ruse, Duncan was a "fellow traveler with the Darwinians", and in a social sense he was. He frequently applauded Huxley's evolutionary palaeontology, acts which did not pass unrewarded (Huxley supplied a testimonial when Duncan applied for the Chair). Duncan's early work was laudatory of Darwin, and he called natural selection "the great excitant".[21] Like many descriptive palaeontologists, he worked extremely close to his material. He showed in 1865 that echinoderms from European Cretaceous deposits and their equivalent in Arabia were not merely specifically identical, but actually had varieties in common. He thus assumed that species could spread widely without changing. But while he determined that many were resistant to change, he noticed that others living alongside were astonishingly variable, almost as though each species had its own "inherent power of variation", quite independent of physical conditions.

Not until after Mivart defined his position (1870–3) did Duncan tentatively shift to a polyphyletic explanation. In his Presidential Address to the Geological Society (1877), he denied the Darwinian premise that unchanging conditions mean unchanging species, and he again looked to "some positive energy" within organisms themselves sufficient to produce "progressive changes". Like Mivart he pointed to the parallel development of marsupials and placentals as proof that evolution was not regulated by "competition or the influence of external conditions alone". Indeed, taking a distinctly Mivartian line, he tackled the problem that had vexed everybody, Huxley included – namely the similarity between living Australian mammals and the marsupials of Mesozoic Europe. Rather than resort to migration, worldwide distribution, or sunken supercontinents, Duncan speculated that "the Triassic Marsupials of Europe have had a direct reptilian genesis". That is, they were born on the spot, while Australian mammals originated independently from egg-laying monotremes, and these from Huxley's bird-like dinosaurs. In accepting more than one point of origin, he suggested, "some of the greatest difficulties in the explanation of the distribution of plants and animals would be removed."[22]

What do these cases prove? First, that Huxley could actively recommend 'progressive evolutionists', Dawkins for Manchester and Duncan for King's; this suggests that he was more or less happy (or resigned to) non-Darwinian fellow-travellers, at least on scientific grounds. The further fact that he provided both with testimonials in 1869, at the time when he was cold-shouldering his "constant reader" Mivart, implies that more than science was at stake in the latter's case. Mivart's attempt to reconcile Rome and Darwin, his irreverent tone and overtly Owenian idealism, provided Huxley with a stick with which to thrash those ritual-bound institutions, Platonism and Catholicism. This explains why Mivart's morphological and anatomical criticisms (many of which, judging from Duncan and Dawkins, were unanswerable) went by the board in Huxley's review of the *Genesis*, while Father Suarez came in for corrective treatment. Neither Duncan nor Dawkins indulged in open Mivartian-style apologetics, being of a quite different disposition (*Nature* was to commend Duncan's "quiet, unostentatious way"[23]): grass-roots palaeontologists who stayed at their posts and puzzled at the scientific problems rather than retreating into a world of "reconciliation". Giving Huxley little excuse to take offence, they were welcomed into the evolutionary establishment like their senior 'progressive' colleagues, Lyell and Carpenter.

Grasping the Divine Idea – Seeley on Pterosaurs

> I was a student of law at a time when Sir Richard Owen was lecturing on Extinct Fossil Reptiles. The skill of the great master . . . taught me that the laws which determine the forms of animals were less understood at that time than the laws which govern the relations of men in their country. The laws of Nature promised a better return of new knowledge for reasonable study. A lecture on Flying Reptiles determined me to attempt to fathom the mysteries which gave new types of life to the Earth and afterwards took them away.
>
> Harry Seeley, now sixty-two and quite deaf, opening his *Dragons of the Air* (1901).[24]

The College of Surgeons was conveniently sited in the "green oasis" of Lincoln's Inn, enabling Owen to recruit a number of law students from the Inns of Court next door – a source, appropriately enough, of some of the most Romantic "legalists of nature" of Victorian times. Unlike Mivart, Harry Seeley (1839–1909) was never called to the Bar, being drawn from law to Owenian studies while still in his teens, but he is of extreme interest for far-outflanking both Owen and Mivart in his idealist palaeontology. He was notoriously uncompromising, and thus from a positivist perspective something of a 'failure' (perhaps explaining his absence from the influential *Lives and Letters* of Darwin, Huxley, *et al*); and his intellectual eclipse was assured by an inhospitable twentieth-century historiography which sifted the 'winners' from the 'losers' rather than relating divergent attitudes to their various milieux. There is no denying that Seeley was "a man of marked individuality, very independent in opinion and original in thought."[25] So much so that his was the perennial dissenting voice at the Geological Society. Never flamboyant, he nonetheless gave sobriety a sort of anarchic tinge. It is impossible to tell what other palaeontologists thought of him. Perhaps only the Philadelphian E. D. Cope, having a certain amount in common (a like of reptiles and dislike of Huxley), singled him out for particular praise. A "fast friend thro' evil as well as good report", Cope was to call him, and "one of the ablest men in the country . . ."

Seeley's early life as a labourer shows that the social elite held no monopoly on transcendental views, but that education could be crucially important. Indeed, one gets the feeling that Seeley's uphill

Harry Seeley. (The author.)

struggle to achieve respectability goes a long way to explain his battling attitude and fierce independence. His father was bankrupt and forced to work as an artisan, having ruined himself "with scientific experiments", and Seeley as a boy was apprenticed to pianoforte makers and lived in a factory.[26] He was acutely conscious of his station, even sensitive – and not without reason; he told Sedgwick that he failed at law school (the refuge of gentlemen) because he was stigmatised as one of the "great unwashed". His childhood had an almost Millerian turbulence; he was nervous and excitable, and prone to collapsing. Indeed, we only know these intimate details because he came close to a breakdown at Cambridge, and to explain his mental state wrote a series of long letters to Sedgwick relating the tragic story of his upbringing.

With so complex a character, it is not immediately apparent why he needed to go to such transcendental extremes – until, that is, we look at his education. From the age of eleven the boy with his father had sampled the various scientific institutions in London, and heard Edward Forbes and later Huxley at the new School of Mines (finding Huxley's lectures the "less attractive"). In the later fifties Seeley joined the Working Men's College in Great Ormond Street – an institution founded by Christian Socialists in 1854 and supported by F. D. Maurice, that clerical bulwark against *laissez faire* and its social iniquities. Here he was "petted and made a good deal of", and met Ruskin and Maurice himself. And he wrote his first paper, on chalk starfishes, in November 1858 while Secretary in the college museum (which he helped found), supporting himself meanwhile by copying manuscripts at the British Museum. Owen's lectures might have given him "the means of dealing with the vast accumulations of vertebrate fossils" he found at Cambridge, but to my mind it was Forbes' influence that stands out most. He taught Seeley that species were variations clustered around "generic ideas"; and by 1861 Seeley was advising would-be geologists of the necessity of "grasping the Divine Idea".[27] In practical terms, this meant laboriously computing all specific variations until the Divine theme dawned – "the one thought in all the varied drapings of the species". From the outset, the young muscular Christian had devised a programme that was profoundly different from Darwin's.

But Cambridge was perhaps the decisive factor – here he believed he could take his stand "not as a working man, but as a man". He went up in 1859 to help the ageing Adam Sedgwick (now 74) catalogue the fossils in the Woodwardian Museum. Cambridge was still a haven of privilege and moral order, and now it was also taking steps to lay the social, political, and scientific spectres raised by industrialisation. Being committed to Established Church and Tory politics, the university's strength lay in a conservative squire-archy with a rural–gentry rather than urban–commercial base, and it naturally found any mercantile intrusion into science or politics abhorrent. So what finally pushed Seeley out on a Platonic limb was probably the Cambridge mood of spiritual reaffirmation: that re-newal of faith in a transcendent reality described by Sheldon Rothblatt in his book *The Revolution of the Dons*. Brian Wynne talks of the university's need for "metaphysical unity to give coherence to a moral universe" imperilled by "atomistic nihilism".[28] Clois-tered in Cambridge, Seeley's social identity began to blur. He still aspired to be a kind of scientific Cobden, travelling the country, lecturing "to every town & school on the practical utility and commercial advantage of science", and yet he began to show a surprising indifference to new London intellectual trends, actually admitting in 1867 that he had never read a word of Spencer, nor did he know anyone who had. Though he moved back to London on Sedgwick's death in 1873, first taking the Chair of Physical Geo-graphy at Bedford College, and in 1876 the Chair of Geography at King's, his science always retained a distinctive flavour, defiantly at odds with progressive metropolitan trends.

Seeley became an acknowledged expert on pterodactyls, and to appreciate the strength of his Platonic science we might look at the extraordinary way he tackled these flying saurians. Before his arrival at Cambridge, Sedgwick had sent his local Greensand fossils to Owen for identification, accompanied by such notes: "What have you made of the lot I sent you last? There was at any rate a *new Pterodactyl*; or a monstrous form of an old one – if we are to throw out *species* into that 'paradise of fools' – 'that limbo large & broad' – now called transmutation."[29] In this anti-evolutionary environ-ment, Seeley was now encouraged to take over identification and description, and Sedgwick personally helped finance publication of the results. As early as 1864 Seeley had struck up a radical position, announcing in the title of his paper to the Cambridge Philosophical Society that "Pterodactyles are not Reptiles, but a new Subclass of Vertebrate animals allied to Birds". He deliberately cocked a

snook at Cuvierian convention, which would make flying reptiles
rather exotic lizard relatives. The hollow limb-bones and vertebrae
he attributed to an avian-like lung-sac system, and he suggested that
with the heart and circulation able to meet the rigours of active
flight the animals "must have had hot blood". He was also able to
expose a cranial cast in one specimen, and wrote triumphantly to
Owen: "The brain is in no respects reptilian; in the main avian, with
a suspicion so to speak of the mammal; altogether a peculiar brain"
raising the animal out of the reptilian grade.[30] Of course, birds and
pterosaurs differed on points, but Seeley – like Cope – saw
Archaeopteryx, with its teeth and tail, begin to bridge the gap. Even
the wings, though they differed crucially from a bird's, were
nonetheless thought to have been covered by fine feather or hair.

The operative point in all this is Seeley's emphasis. He was not
making an evolutionary statement. His palaeontology made only
minimal contact with the "derivative" world, and he shunned
phylogeny or the tracing of family trees. His arguments were solely
about *rank*, his criterion was "the principle of organization", and his
solution was to give pterosaurs parity with birds on account of their
similar blueprint. This insistence on immutable plans behind the
façade of nature far outshone Owen's or Mivart's, at least in
practical terms. For them, at any rate, the Word was *progressively*
made Flesh, and "derivative creation" was a legitimate subject of
study. Thus in the massive *Anatomy of Vertebrates* (1866–8) Owen
intimated that pterosaurs were an "important link" between birds
and reptiles,[31] and he gave this link a derivative twist a few years
later in his brushes with Huxley.

Actually Owen refused to see pterosaurs as anything but cold-
blooded reptiles, stamped with little that was specifically avian. We
saw in Chapter 4 how Huxley in 1867 also speculated on the
avian-like anatomy and physiology of pterosaurs and dinosaurs (to
bolster his projected dinosaur-bird relationship). Owen now responded by slapping down Huxley and Seeley in turn. Lack of
insulation (and the entombing sediments were often fine enough to
record hair or feathers had they existed) and a "miserably small"
braincase convinced him that these were truly reptilian. Seeley
protested in the *Annals and Magazine of Natural History* (1870) that
Owen was "erroneous, unscientific, and unjust", and he met with
some support, for his sentiment, if not his science. Perhaps the most
contentious point concerned the pneumatic bones: these Owen

dismissed as "purely adaptive" weight-reducing characters, as "advantageous to a cold-blooded as to a warm-blooded" flyer.[32] But the concept of the "adaptive" character could serve many masters and subtly change its meaning with each one (a graphic instance of underlying ideologies colouring the meaning of words). Thus Huxley also considered the hollow bones "merely adaptive modifications",[33] by which he meant that they *were* physiologically important. At the time he was trying to nip the idea of a pterosaur-bird relationship in the bud (before introducing his dinosaur line of descent). According to him, the air-filled bones in birds and pterosaurs were not part of a common birthright, as Seeley maintained, but produced independently in response to similar flight-pressures. And this belief sent Huxley as unswervingly towards "hot blood" in pterosaurs as Owen was led away from it.

So even on such esoteric matters conflicting interpretations could be generated by prior theoretical considerations, themselves grounded in distinct social strategies. Seeley's ranking of fossil life reflected the cosmic moral and social order he encountered at Cambridge. Huxley was keen to prove the physiological sophistication of some reptiles to further his evolutionary speculations (themselves rooted in the bourgeois ideal of self-betterment). And Owen was attempting to preserve the status quo in face of the positivist threat. The connection with pterosaur physiology might seem tenuous, but one can reconstruct a definite relationship between such deeply-rooted social commitments and the more abstruse palaeontological consequences.

Seeley's *Ornithosauria* (i.e. "Bird Reptiles") was published to mixed reviews in 1870, and the *Annals* critic for one thought his attempt to upgrade pterosaurs somewhat shaky. But the book was more than a bid to "determine the Pterodactyle's place in nature".[34] It was a radical attempt to shake up preconceived notions about saurian life; a caution against squeezing extinct reptiles into preconceived moulds, conditioned by living lizards and their allies. Owen more than any one had highlighted the *spread* of the Reptilia, from fish-like archegosaurs at one end to mighty dinosaurs anticipating the mammalian grade at the other. Seeley agreed that the influx of exotic ichthyosaurs, plesiosaurs, dinosaurs, pterosaurs and anomodonts, like the South African two-tusker *Dicynodon*, had rendered the "Reptilia of the Palaeontologist" a "vast and provisional group". He was rightly worried lest we straitjacket these extinct groups. But rather than trace out lineages by hunting the ancestors of the "higher" forms among the "lower" – as Huxley and Owen in their ways were doing – his solution was to hive off groups and locate them elsewhere. He doubted whether dinosaurs or ichthyosaurs really were reptilian. Had we been able to infer soft structures, as he had done for pterosaurs, we should probably find them constructed according to somewhat different specifications. This approach was ultimately inimical to evolutionary palaeontology; at least, Seeley seemed forever intent upon bringing down bridges as soon as they were raised. Answering Huxley's case for an avian ancestry, for example, he insisted that the long hind legs of some dinosaurs were "due to the functions they performed rather than any actual affinity with birds". Then he effectively removed dinosaurs from contemporary reach by insisting that they were as "unlike reptiles as birds or mammals."

If his palaeontology was hopelessly at variance with the phylogenetic approach pioneered by Haeckel, it did have an inner Platonic logic. As the Cambridge intellectuals re-emphasised the moral order of the universe, so Seeley produced a palaeontological parallel: "unity is incontestable", he wrote, and no superficial change, "whether it is named creation or whether it is named evolution" can disguise the fact that behind the transient stands the eternal. And however change is envisaged, it can only lead to our "unutterable and reverent confidence" that all life is transcendentally planned:

> For me it indicates, beyond laws and their consequences, what, judged by human standards, is Intelligence, of which laws in their working are manifestations. If, then, an attempt is made to explain the plans of animal life, it is in faith, born of science, that they are the products of divine law . . .[35]

Seeley assumed that the whole point of doing science was to reveal the enduring Thought, the transcendent reality, not chronicle shadows flitting across the globe. In search of this deeper reality, he took up a problem dropped by Owen a generation earlier. He tackled the vertebral origin of skull and limbs, in what must have looked to evolutionists like a continued quest for the holy grail, the ultimate 'unit' or organisation. According to modern historians, Huxley had already shown the futility of this in his Croonian lecture of 1858, when he traced the development of the skull in the embryo. Nothing is so certain if Seeley is to be believed. For him, as for Cope, the issue of Owen's 'vertebral' skull was still alive and kicking, and he produced three meaty papers (1866–1882) which paid lip service to Huxley's embryological refutation while praising Owen's "beautiful" descriptions. Seeley concluded that the "brain-case", at least, was "a modified vertebra" and cheerfully predicted that embryology will one day become "the servant instead of the lawgiver of morphology" (meaning that Huxley's position must yield to Owen's).

The socially-stratified moral cosmos of the Cambridge dons found its microscopic reflection in Seeley's brand of Owenism, with its emphasis on order and rank rather than phylogenetic (= social) mobility. To the extent that Cambridge tried to maintain standards by becoming politically and socially repressive, so Seeley – in snatching pterosaurs and dinosaurs from the evolutionists and placing them beyond reach – seemed insensitive to the new palaeontological needs (or rather, those of self-confessed "plebeians" like Huxley). On the other hand, it is difficult to deny that his obsession with 'plans' and 'grades' paid handsome dividends. It prompted tantalising speculations of pterosaurian physiology, and probably had spin-offs in classification. For example, in 1887 he split the dinosaurs into "lizard-hipped" and "bird-hipped" forms on the strength of consistent differences in pelvic structure, rather than any study of phylogenetic trends.[36] It stemmed more from his love of morphological tabulation than any evolutionary imperative. So there is no despising Seeley for being a metaphysician in an age of strong anti-metaphysical sentiment, and one has to admire a man who can produce a science so exquisitely adapted to his social needs. Moreover, anti-Platonists like Huxley who ignored "grades" lost a wonderful explanatory tool and – as we shall now see – occasionally became mired as a consequence.

Why Did Huxley Ignore the Mammal-Like Reptiles?

Until the 1970s Huxley's science was rarely treated with anything like critical detachment. For toppling the unpopular Owen he has always been loudly hailed. No one can doubt the one-time tactical advantage in reslaying the slain, or despise older evolutionist-historians for continuing to deride Owen's ideology, which they portrayed as totally unconducive to 'good' science. But these days we need to take a balanced, sympathetic stand, and should be more interested in the production of knowledge than the destruction of straw-men. Just as Seeley's and Mivart's philosophies were productive in their context, so Owen's had an enormous heuristic value. Even late in the Victorian "afternoon", his ideology, 'bias', or what you will, led to his successful solution to the question of mammal origins – one area in which Huxley was particularly weak. In fact their contrasting attitudes to the ancestry of mammals makes a salutary study.

Consider Huxley's position first. In 1879 he published a seminal paper in the Royal Society's *Proceedings* on mammalian anatomy, which ended with some "Conclusions respecting the Origin of Mammals". The paper was eminently theoretical and curiously avoided fossils on the whole (despite his former insistence that the evolutionist must produce the "title-deeds" to his scientific estate – that is, fossils to back his case). Instead he compared the primitive platypus, living reptiles, and the salamander. Internally, his study made marvellous sense. He was able to prove that mammals could not possibly have descended from the reptile-bird stock, only from the amphibians, presumably via "some *unknown* 'promammalian' group" (Haeckel's term).[37] There was never any dispute of the basis of Huxley's conclusions. The layout of the aortic arches, for example, is so different in birds and mammals that the divergence must have occurred in extremely archaic stock (whether amphibians or something later is a moot point). Huxley visualised twin evolutionary streams, one stretching from amphibians through reptiles and dinosaurs to birds, the other departing at the amphibian level to reach mammals through some undiscovered "promammalian" group. So the amphibian ancestry was perfectly reasonable in its way, but the point I wish to emphasise is his use of "unknown" intermediates. He wrote:

> The discovery of the intermediate links between Reptilia and Aves, among extinct forms of life, gives every ground for hoping that, before long, the transition between the lowest Mammalia at present known and the simpler Vertebrata may be similarly traced.

This can only mean that they had not already been discovered, and as if to explain Huxley's recourse to amphibians as ancestors Bowler wrote that only "toward the end of the century did the Karoo formation of South Africa begin to reveal the existence of a series of what came to be known as the 'mammal-like reptiles', described by workers such as Harry Govier Seeley."[38] Were this the whole truth there would be no problem – we could understand why Huxley was still casting around for mammal-ancestors in 1879.

Seeley did begin a serious study of the mammal-like reptiles in the late 1880s, travelling to South Africa for the purpose. But by that time Owen had been discussing the self-same fossils for three decades. In fact, according to his son-in-law (the palaeontologist Arthur Smith Woodward) Seeley actually *followed* Owen's work for many years, and only after Owen retired did he decide to extend it, as though it were a policy decision on his part. So the question at the head might better be put: 'Why did Huxley side-step *Owen's* mammal-like reptiles?', since these should have made suitable mammalian ancestors.

The story of these contentious fossils is evidently poorly known. So to establish that there *was* a wealth of fossil material for Huxley to choose from, we might start from scratch, using a number of unpublished letters to give a fuller picture. Discovery of the Karoo reptiles accompanied the colonial expansion in South Africa. The first "two-tusker" *Dicynodon* reached the Geological Society in 1844, shipped by a military road builder, Andrew Bain. The Council took an immediate initiative and voted him funds, and with perfect diplomacy the President, Henry Warburton, wrote that Bain had "taken a most judicious course in sending them [the fossils] to our Society."[39] Bain was overjoyed at the size of the reward, swankily admitting that he had received the "highest honor yet obtained by any Cape Colonist, viz. a present of £200 from the Royal Bounty Fund . . ." The immediate result – he shipped his entire collection to the Society. By 1851 he had also laboured for fourteen years on the first geological map and sections of South Africa, and he now prepared to lay the fruits before the Council, to support his contention that the continental interior had once been "an immense fresh water lake!"[40] His letters home made it clear that "South Africa is richer in fossil remains than it was ever thought to be . . ." Unfortunately, their preservation was often poor. Exposed specimens were badly eroded (a situation not helped by the

Boers, who were given to smashing them), and those still in matrix were often impossible to extricate. Bain himself complained at the outset that the sandstone was exceptionally hard, and "so amalgamated with the bone that it was not only difficult to separate the one from the other, but even to distinguish which was which." *Dicynodon* required expert handling from the masons at the British Museum. And this is where many of the specimens tended to end up, partly because of the Museum's facilities, but also because – after Owen's arrival in 1856 – the Trustees took over Bain's remuneration and effectively diverted supplies.

Not only paid collectors, but a whole field network, from the Governor to missionaries and explorers, helped locate and (via the provincial geological societies) ship fossils back to Owen. Local geologists had even tried to co-opt the Dutch Calvinist farmers into helping, Seeley was told, "but stern puritans that they are, with one exception, they gave it up, when they saw that its conclusions tended to undermine their primitive creed as to creation and the age of the world . . ."[41] One of the Royal Children, Prince Alfred, fortifying the British presence in South Africa with a visit in 1860, had actually returned armed with two dicynodonts for Owen. Even as private tutor at the Palace, he could hardly have expected so practical a reward. Of the greatest importance were a batch of fossils still in their sandstone matrix dispatched by the Governor of the Colony, Sir George Grey – two of which Owen singled out at the Geological Society in 1859 as unique. One he called *Galesaurus*, or the "weasel reptile", on account of its low, flat skull, and the other *Cynochampsa* or "dog crocodile". Both possessed something unheard of in reptiles: a mammal-like dentition, divided into incisors, stabbing canines, and molars. Owen never doubted that these were reptilian skulls. Yet he was obviously struck by this 'canine' dentition, in which the Karoo predators "made a most singular and suggestive approach to the mammalian class".

But the most consistent and reliable suppliers were still the Bains, Andrew, and after his death, son Thomas (who was later to escort Seeley round the sites). Between them they ensured a continual stream of fossils right through the century, and Owen was able to report in 1876 that "specimens have reached the British Museum almost year by year to the present time." They amounted to a good half-dozen genera and several species, enough in fact to warrant a *Descriptive and Illustrated Catalogue* (1876). With names like *Tigri-*

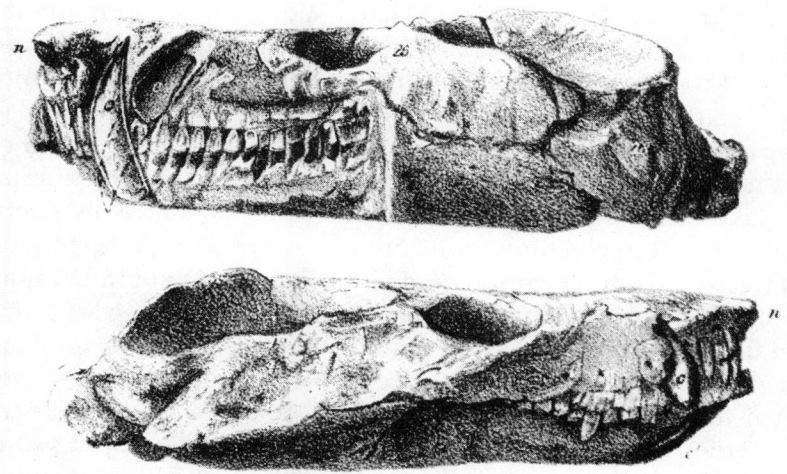

The South African mammal-like reptile *Galesaurus*. (From Owen [1860c].)

suchus ("tiger crocodile"), *Cynodracon* ("dog dragon"), *Lycosaurus* ("wolf reptile") and *Cynosuchus* ("dog crocodile") no doubt was left where analogies were to be sought. To house them, Owen at this time erected a new order Theriodontia ("beast tooth"),[42] a move designed to capture the spotlight. (Although Seeley of course was dubious at first, imagining this to be a rag-bag of unrelated types.)

From the late seventies, Thomas Bain was supported by Treasury grants (on Owen's advice), both to reimburse him and finance future ventures. Hence he was voted £200 for the first full-scale expedition to Beaufort by ox cart in 1877. Unfortunately it was also a time of severe drought, when he had barely enough water to keep his men digging, and he was forced to leave many specimens. All the same, he came away with "almost 280 heads",[43] by no means a bad haul. Not surprisingly, when Seeley eventually arrived in the Colony in the summer of 1889, he discovered "that most of the important specimens collected had already been sent to Sir Richard Owen . . ."

So the evidence is pretty conclusive. Not only were the 'mammal-like reptiles' shipped to Britain in great quantity during the seventies, but more importantly, they fell under Owen's jurisdiction. Officially, of course, as British Museum property, they were open to inspection, but one can imagine Huxley's difficulty or even embarrassment had he tried to see them. Consider the way Owen treated 'his' fossils, according to Henry Woodward, Owen's junior colleague at the Museum for a quarter of a century. Woodward claimed to "stand in a more intimate and personal relation to him than many others",[44] but was no less irritated by Owen's quirks. Apparently his "whole nature seemed transformed in the presence of a new and undescribed fossil", and he was overcome by a desire "to take the field and describe it". If however "the new discovery arrived at an unpropitious moment, he was equally eager to conceal his treasure from the curious and inquiring eyes of youthful aspirants." Nothing could have been more galling to a young assistant, and it helps explain the awful aura of mistrust which surrounded Owen. Of course stealth and secrecy are not historical prerogatives when it comes to such priceless relics, at times they appear an occupational hazard. Nevertheless, the nature of the subject seems to have brought out the worst in Owen, which is one possible reason why Huxley – though he often described the Museum's fossils – generally left the South African theriodonts alone.

For all that, the 1876 *Catalogue* should have been the answer to Huxley's prayers. It described a dozen carnivores which, in Owen's view, nicely filled the gap between mammals and reptiles. Consider one in particular: *Cynodracon major* was a formidable predator about the size of a lion, with a flexible fore-paw for "seizing and lacerating" its prey. Its teeth resembled those of a sabre-tooth cat – so much so, said Owen, that were only the teeth known, no one would have doubted that big cats had lived in Triassic times. These were reptiles and he listed numerous details as proof, adding in 1881 the lack of milk teeth in the young. Nonetheless, in a "Summary" at the end of the *Catalogue* he again emphasised the mammalian features of many reptiles, linking the South African genera in particular to marsupials and lowly placentals. He reiterated the skull and skeletal similarities, the paw, and the stabbing teeth, such feline features previously being "utterly unknown and unsuspected as reptilian ones". He pointed out that today's lizards and tortoises give no clue to the great "gains in organization" the ancient reptiles had once made. Owen was not one for sweeping evolutionary generalisations; he was always reticent and his language guarded. But he does go on to say that these gains have "been handed on, continued, and advanced, through a higher type of Vertebrates".

Put another way, the enormous strides made by Triassic reptiles had been built on by early mammals – although exactly how this was achieved was "unintelligible to the writer on either the Lamarckian or the Darwinian hypothesis."[45] But that did not stop John Evans welcoming Owen's "highly important paper". It gave "no uncertain sound", he said, nodding in Flower's direction, "as to the probable affinities between some mammalian and reptilian forms", and he placed it alongside the discoveries of Gaudry, Kovalevskii, Marsh, Leidy, and Huxley, in proving that all present day vertebrates are "direct descendants of those of earlier periods."

In the five years between the *Catalogue* and his description of an ageing individual *Aelurosaurus* ("cat reptile") in 1881, Owen published a string of papers reinforcing his position – although even in the 1880s he had to admit that the original weasel-like *Galesaurus* still had "the advantage over all the subsequently discovered Theriodonts in the entireness of the skull."[46] So why *did* Huxley in his 1879 paper on mammal origins neglect to mention Owen's "ancient Triassic precursors of our existing cats"? A fine study of Huxley's psychology can probably provide no more than a partial answer. There is no question that he was wilfully suppressing evidence, or, having been effectively barred from examining the Karoo imports, that he was wreaking some kind of vengeance (blackballing Owen's science would have proved self-defeating, if nothing else). It was once said that "no man ever manifested more of the moral presuppositions of a Puritan evangelism",[47] and since morality for Huxley rested in acquiescence to 'positive' Truth, he could no more deny a 'fact', Owen's or otherwise, than believe in the immaculate conception: it would have been professional suicide. (For good reason were his lectures collected under the title *Lay Sermons*.) Therefore he was strongly persuaded of the irrelevance of Owen's fossils, and must have put up good scientific justification.

A promising explanation, at first sight, is his faith in the 'persistence' of mammals, the relic of his Lyellian strategem against the progressionism of Owen and Chambers. In 1869 – as we saw in Chapter 3 – he was still convinced that mammals had evolved by Silurian times, in which case they could hardly have been born of Triassic parents – and this would have ruled out Owen's reptiles as the immediate ancestors. Although the seventies saw Huxley's neo-Lyellian views mellow, as he felt the growing need to underpin Darwinism with an evolutionary palaeontology, in 1879 he still

Richard Owen in old age. (By permission of University College London Library.)

sought his "promammals" in the "late Palaeozoic epochs".[48] However, his belief in the early evolution of *birds* did not prevent him from formulating a dinosaur ancestry – even though known fossils, the Jurassic *Archaeopteryx* and *Compsognathus*, were much too recent to have been directly ancestral. He rationalised *this* evolutionary route in *American Addresses* (1877) by conceding that the bird-like dinosaurs were therefore the "more or less modified descendants of Palaeozoic forms through which the transition was actually effected."[49] So here he retained the best of both worlds: his faith in 'persistence' and a Palaeozoic origin for birds remained intact, yet he managed to use Mesozoic fossils as evidence of the evolutionary route. Nothing stopped him from applying this argument, *mutatis mutandis*, to mammals, making Owen's theriodonts the Triassic survivors of the ancient transitional stock, something that Marsh's Bohemian assistant Georg Baur was certainly prepared to do.

Ostensibly, it was a purely scientific criterion which stopped Huxley short, namely the impossibility of deriving a mammal's aortic arches from a living lizard's. Yet it was apparent to the Cope-Owen-Seeley faction that there was something wrong with his logic; or rather, that it did not apply to certain *extinct* reptiles. Cope was by now demanding a central role for his mammal-like reptiles, having described the large crested *Dimetrodon* from the Texas Permian in 1878. He went on to ally the Texas and Cape fossils in the order Theromorpha ("beast form") and in 1880 made it easy for Huxley by pointing out that *Dimetrodon*'s relatives were remarkably similar in some respects to the Texas Permian amphibians. Moreover, the long-cherished belief that Triassic and Jurassic mammals were necessarily marsupial was finally under siege. Seeley in 1879 actually went so far as to create the novel concept of a "generalized order" to house Mesozoic mammals (something presumably Huxley could *never* condone). Within a year Marsh had followed suit, describing the sixty or so *Dryolestes* specimens collected from Como quarry in Wyoming as "manifestly low generalized forms, without any distinctive Marsupial characters",[50] and he too erected for the majority of Mesozoic mammals a new 'sub-marsupial' order (which he called the Pantotheria).

So between them Marsh, Cope, Owen, and Seeley laid the foundation for the other evolutionary stream – from amphibians through the mammal-like reptiles to the "generalized" Mesozoic mammals and thence modern marsupials. Twentieth-century palaeontologists had little difficulty visualising reptiles as a *grade* rather than a class – accepting that as dinosaurs passed through the grade on their way to birds, so theriodonts did on their way to early mammals. Such an option was probably denied Huxley. His life-long struggle with the idealists, and the inextricable development of his positivist and anti-"realist" philosophy, presumably left him considering "grades" of Platonic imposition. Indeed, with Seeley himself urging the use of "Fossil Reptilia" to fill the "morphological interval between Amphibians and Mammals" on the one hand, and "reptiles and Birds" on the other,[51] Huxley may well have been right.

Conclusion

If not only abstract concepts (phylogeny, persistence, progression) but even the more 'factual' aspects of palaeontology *do* have a constitutive ideological dimension, then it is surely less profitable to judge a theory 'right' or 'wrong' using hindsight than to see it as an adaptation to a specific context. (And since social milieux change, evaluating the past on the present's terms is little less than arrogant.) We should therefore transcend the impoverished historiography which cheers Huxley for vanquishing religious 'obscurantism' or dispatching the chimerical archetype. Better that we treat the Huxley-Owen debate as tangible evidence of a more fundamental ideological divide – while freely admitting that the ultimate socioeconomic 'bases', if such there were, lay well below the palaeontological 'superstructure', and were mediated at innumerable levels by religious, political, and class considerations.[52] It would be an oversimplification to say that Huxley and Owen spearheaded rival factions on opposite sides of the Industrial divide. Nonetheless, it is profitable to view the two sides manoeuvring for power within the scientific community – the young "plebeian" materialists, whose attack on God-given rights gained them "professional" autonomy, vying with the largely establishment- and church-supported idealists, whose primary allegiance was reflected in their universe of moral and social order. With such a deep ideological divide, the crucial debates over human uniqueness obviously ended in a stand-off, with neither side retreating to any significant extent. Despite a positivist historiography which harps on Owen's 'defeat', it comes as no surprise that idealists like Mivart and Argyll continued to congratulate him in post-hippocampus times for having the right intent.

Scientists do not work in a vacuum, but tend to reflect the views of the larger culture. Hence in Darwin's day, the idealist package of polyphyly, taxonomic ranking, and purposive evolution was well adapted to an ordered, privileged existence – one threatened by the bourgeois tree of life which promised to reduce all men alike to lowly parentage. The social stakes in the wake of industrialisation were high; as a result the concepts of palaeontology were often of crucial importance. Not for nothing did Owen's 'pachydermal' and anti-transmutatory dinosaur stand implacably opposed to Huxley's delicately-avian and evolutionary one. However, rather than championing the 'winner' Huxley – the prototype of the twentieth-century biologist – as many partisan studies do, we should realise that *both* men made lasting contributions. Huxley's vested interest resulted in spectacular advances in our knowledge of avian evolution. And Owen's commitment to the 'mammalness' of dinosaurs undoubtedly explains his success with the Karoo theriodonts. [53]

And successful he was. While Huxley abortively searched for his hypothetical "promammals", Owen was unloading crates of fossils from the Cape which conveniently fitted the bill. In 1880 he finally announced that the platypus was a "remnant" of Karoo stock, and that the marsupials had "branched" off at some point. [54] This, a Whig would say, was near enough 'right' – the truth was out. But it is clear from a contextual point of view that the 'truth' had more to do with social pressures than scientific prophecy.

Notes and References

Abbreviations used in citing sources of manuscript material.

BM(NH) OC	British Museum (Natural History), Owen Correspondence
IC	Imperial College of Science and Technology, Huxley Papers
BL	British Library
RCS	Royal College of Surgeons
UCL	University College London
CUL	Cambridge University Library

Introduction

1 Quoted in Geoffrey Best's *Mid-Victorian Britain* (1979), 27; see also 25–7, 76.
2 Young's *Portrait of an Age* has recently been reissued as an annotated edition by Kitson Clark, G. M. Young (1977), 92.
3 G. K. Clark (1962), 30, also 31–3.
4 Burn (1965), 71.
5 Barnes & Shapin (1979), 92. Of related interest are Turner's essays (1978 and 1980) and Chapter 2 of his *Between Science and Religion* (1974); and also Roy Porter's "Gentlemen and Geology" (1978), with its sensitive critique of the "Professionalisation" model and recognition of a prior "career" stage in nineteenth-century geology. He discusses "professionalisation" at the Geological Survey and the threat posed to "gentlemanly geology" by increasing specialisation after mid-century, and goes some way to explain the growing conflict between "the elite

and the weekend amateur" (830) – on which subject see Allen (1978a), 191.

6 Becker (1874), 248.

7 Flett (1937), 55; Porter (1978), 833.

8 Owen (1878), 428–9, on which see my Chapter 5. Becker (1874), 48–9, mentions the loss of palaeontology's appeal.

9 F. Darwin & Seward (1903), II, 13; Cope referred to London as "Babylon" in a letter home to his wife in October 1878: Osborn (1931), 253.

10 A. Von Kölliker to Huxley, 10 January 1860, IC 19.282 (cf. his 1858 letter, 19.280).

11 Some important recent books dealing in whole or in part with nineteenth century palaeontology in Britain are Dov Ospovat, *The Development of Darwin's Theory* (1981); William A. S. Sarjeant, *Geologists and the History of Geology* (1980), a monumental five-volume bibliographic survey; L. J. Jordanova & Roy Porter (eds), *Images of the Earth* (1979); Michael Ruse, *The Darwinian Revolution* (1979); Stephen Jay Gould, *Ontogeny and Phylogeny* (1977), esp. Part I, pp. 13–206; Roy Porter, *The Making of Geology* (1977), which stops at c. 1815; Peter J. Bowler, *Fossils and Progress* (1976), an important internalist study; M. J. S. Rudwick, *The Meaning of Fossils* (1972), already a standard work; and C. C. Gillispie's classic *Genesis and Geology* (1959), still readable and immensely valuable.

12 On the paradoxes and problems of Whiggism, see Oldroyd's essay (1980b).

13 Cornish (1904), 38.

14 Jensen (1970), 63, quoting from Thomas Archer Hirst's Journal, see also MacLeod (1970).

15 W. B. Carpenter to R. Owen, 23 September 1842, BM(NH) OC 6.308.

1. Huxley, Owen, and the Archetype

1 Foster (1869), 382; Nordenskiöld (1942), 415, calls Owen "England's greatest comparative anatomist".

2 Owen & Broderip (1853), 83; L. Huxley (1900), I, 93–4.

3 Rev. R. Owen to T. H. Huxley, 25 September 1895, IC 23.251; Huxley replied on 26 September 1893, IC 23.253; and related the incident to J. D. Hooker on 1 October, IC 2.429, where he uses five exclamation marks to show his astonishment and talks of "poor old Richard's ghost" (for Hooker's reply see IC 3.408). See also L. Huxley (1900), II, 364, 373; and Huxley (1894) for the chapter in question.

4 MacLeod (1965), 280; Irvine (1959), 38; de Beer (1964), 165.
5 L. Huxley (1900), I, 142 – he was, of course, referring to Huxley's action over Owen's assumed "Professorship" at the School of Mines.
6 Ruse (1979), 142–4; Ghiselin (1980), 107–8.
7 T. H. Huxley to Edward Forbes, 27 November 1852, IC 16.72.
8 L. Huxley (1900), I, 95; Rev. R. Owen (1894), I, 197.
9 L. Huxley (1900), I, 68; Flower (1898), 368; Best (1979), 101–111, discusses the scale of middle class incomes in the period.
10 T. H. Huxley to W. H. Flower, 9 January 1894, IC 16.133; Flower (1898), 365, 368; on Owen's transfer to the British Museum, see also Stearn (1981), Chapter 4, and Gunther (1980), Chapters 11 and 12. Huxley reports Owen's attempts to move to the British Museum in 1851 in L. Huxley (1900), I, 93.
11 Owen (1851), 448. He virtually accused Lyell of being blinded by *a priori* considerations (424) and regretted "that the Philosopher should have been suffered to subside so far into the Advocate" (438). Lyell wrote a long and in places sarcastic reply: C. Lyell to R. Owen, 9 October 1851, BM(NH) OC 18.172. In his *Journal*, Mantell put Owen's "violent attack" down to "spite because Lyell had recommended Mr Waterhouse!" Adding after speaking to Lyell on 19 October 1851 that "he will now understand the real character of Owen!": Curwen (1940), 275.
12 I thank Mike Benton for showing me his MS "Progressionism in the 1850s" (1982: in press). My reading of the Mantell correspondence tends to substantiate Benton's thesis. For example, Mantell wrote direct to Patrick Duff (the discoverer of *Telerpeton*) on reading Owen's *Literary Gazette* notice, asking if it was true that Owen had been sent a drawing, followed by the fossil with a request to notice it (all of which Owen claimed). Duff replied that this *was* "substantially correct"; however, on learning of Mantell's interest, Duff told his brother in London to show the fossil

> to any of the leading Geologists in London that might desire to see it. Dr D[uff] wrote me on the 2d Instant that you, Sir C. Lyell and Prof. Owen were seeking to have possession of it and that I *must decide* who was to have priority. I immediately wrote him that if he had not promised it to Professor Owen that you should have the first turn of it . . . (P. Duff to G. Mantell, 24 December 1851: Alexander Turnbull Library, Wellington, MS papers 83:100; see also Mantell's letter, 20 December 1851, ibid.; and Duff's letter to Owen, 29 December 1851, BM[NH] OC 10.208.)

The lampoon which accused Owen of "prigging" Mantell's bones and worrying him to death was "A Sad Case" in *Public Opinion*: Anon.

(1863), 497. For Mantell's *Journal* record of the affair, see Curwen (1940), 279–81; on his spinal injury in 1841, 148–9; his suffering and expectation of a painful and lingering death, 163, 165, 251, 263, 264, 265.

13 *The Literary Gazette* obituary dismissed Mantell as a "diffuser of geological knowledge" and harped on his "weaknesses" and want of exact knowledge: ?Owen (1852), 842. William Hopkins, about to relinquish the Presidency of the Geological Society, had considered Mantell "a man of rare merit" who had "pursued science & successfully for its own sake under difficulties that would have damped the ardour of all but a few". After reading the obituary, Hopkins wrote to Horner: "Have you seen the article in the Literary Gazette of last Saturday? I think it is palpably from Lincoln's Inn Fields. It bespeaks a lamentable coldness of heart in the writer. How sad it is to see great genius combined with such a want of generous feeling" (W. Hopkins to L. Horner, 17 November 1852, IC 18.228). By this time Owen's name was up for adoption as President, and on scientific grounds he did look the best choice. 'Officially' he "declined" the post (Rev. R. Owen [1894], I, 394; Huddleston [1893], 48), but the unpublished correspondence suggests that there was more to it. Science had become a secondary consideration. "I am perfectly convinced", Hopkins explained to Forbes a fortnight later,

> that the Society would be very divided as to Owen's eligibility at the present moment and indeed I should feel it scarcely respectful to the memory of poor Mantell to nominate one whose occupancy of the Chair would have driven him so entirely from the Society had he been living. I shall invite Owen to be a Vice Pres: & hope he will accept the Office and come more among us again. (W. Hopkins to E. Forbes, 4 December 1852, IC 18.224; but cf. Forbes to Owen, in Rev. R. Owen [1894], I, 394.)

In the event Forbes took the chair and Owen a Vice Presidency. He never was to become President.

14 L. Huxley (1900), I, 97.

15 E. Forbes to Huxley, 16 November 1852, IC 16.170, where Forbes also discusses the importance of Owen's word "in all Government matters". The following day Huxley wrote to remind Owen (IC 23.247), and gave his account of the affair to Forbes on the 27th "I wrote to him as you advised . . ." (IC 16.172). To which Forbes replied "He is certainly one of the oddest beings . . ." on 2 December (IC 16.174). The letter of the 27th was badly transcribed by Irvine (1959), 38–9.

16 The long history of friction between Huxley and the Admiralty can be pieced together from his correspondence. He was indebted to Owen

for his appointment as "Additional Assistant Surgeon" of the *Fisguard* at Woolwich, on half pay and with an initial six month leave of absence (Huxley to W. S. Macleay, 9 November 1851, IC 30.3, which contains a detailed account to that date). Indeed Owen was reassured by John Parker on their Lordships' behalf on the 29 November that Huxley *had* been appointed (BM[NH] OC 21.135). Huxley requested Royal Society funding to publish "a series of drawings of, and researches upon, the structure of the Polypes and Acalepha, made on board HMS Rattlesnake . . ." (26 May 1851, IC 30.2; also E. Sabine to Huxley, 9 October 1851, IC 26.4). Huxley went direct to the Duke of Northumberland (First Lord) on 30 March 1852 (IC 30.10). For his continuing struggle through 1853 see Huxley to his sister, 22 April 1853, IC 31.21; E. Sabine to Huxley, 30 October 1853, IC 26.6, and his pestering of the Admiralty for continual extensions of leave until finally being struck off in 1854: IC 30.13–27. Some of these letters are partially published in Huxley's *Life*: L. Huxley (1900), I, 55–107 *passim*.

17 Owen's testimonial for Huxley's application to Aberdeen: 30 October 1852, IC 23.245; and to Toronto, 30 September 1851, IC 23.244.

18 L. Huxley (1900), I, 81, 100, 107; Paradis (1978), 14. On his mother's death and father's imbecility: Huxley to his sister, 17 April 1852, IC 30.17 and L. Huxley (1900), I, 99.

19 L. Huxley (1900), I, 97; Bibby (1972), 25, apparently misread this, as did Ruse (1979), 143. For the trespasser warning "no poachers" and honour healing no wounds, L. Huxley (1900), I, 97, 107. Irvine (1959), 38, calls Owen "infamous to the backstairs of science".

20 E. Sabine to Huxley, 30 October 1853, IC 26.6; Rev. R. Owen (1894), I, 351.

21 Owen to Huxley, 15 March 1853, IC 23.249; Edward Forbes and Thomas Bell refereed Huxley's paper (information courtesy of Mr N. H. Robinson, Librarian to the Royal Society).

22 John Chapman to Owen, 13 January 1848, BM(NH) OC 7.26.

23 Sedgwick quoted in Napier (1879), 492; Sedgwick (1845), 66; Chambers (1844), 233, Rev. R. Owen (1894), I, 255. See also Sedgwick's almost hysterical outburst to Lyell in Clark & Hughes (1890), II, 83; and Chambers' reply to his critics (1846). Milton Millhauser's *Just Before Darwin* (1959) is the only full-length treatment of Chambers; Gillispie devotes a chapter in *Genesis and Geology* (1959), Chapter 6; although by far the most impressive analysis of Chambers' biology to date is Hodge (1972).

24 Forbes (1844), 265; on the growing scepticism among the upper-middle classes and the "dissolvent literature", see Burn (1965), 274. In his *Autobiography* (1904), II, 33, Spencer mentions Chapman's willing-

ness to commit himself to rationalistic books; here too he recalls first meeting Chapman's circle, e.g. Lewes (I, 347–8) and George Eliot (I, 394–399; also Peel [1971], 13) and Owen's support on the abolition of price-fixing (I, 392–3). The Chapman–Huxley correspondence is 12 August 1853, IC 12.168, 23 October 1853, IC 12. 169, and 26 October 1853, IC 12.170.

25 Journals of Thomas Archer Hirst, Vol. I 1847–1850 (entry for 14 August 1848) held at the Royal Institution; F. Darwin (1887), II, 312. Murphy deals with the "Ethical Revolt" against Christianity (1955), concentrating on Newman, Froude, and Eliot.

26 Richard Owen, "Hunterian Lectures on the Nervous System 1842: Lecture I, 5 April 1842", *Manuscripts, Notes, and Synopses. 1842–8*, BM(NH) OC 38.

27 Quoted in Brooke (1977b), 142; the pleas for a repudiation are to be found in Rev. R. Owen (1894), I, 252–5; Chapman asked permission to quote Owen in Chapman to Owen, 13 January 1848, BM(NH) OC 7.26. Chapman was the "correspondent" mentioned in Owen's *Life*, I, 309.

28 Brooke (1977b), 134. A pertinent question, in judging Owen's response, is whether he knew the author personally. Chambers had gone to inordinate lengths to conceal his identity, using a Manchester-based intermediary to channel each revision (transcribed in his wife's hand) to Churchill in London. He was officially exposed in 1884, after *Vestiges* had sold twenty eight thousand copies in twelve editions (Chambers [1884], introduction). I say 'officially' because his cover *was* broken in 1854 when an erstwhile proof-reader made a clean breast of his part in the affair. David Page admitted his complicity in a provincial lecture reported by the *Dundee, Perth and Cupar Advertiser*, 24 November 1854. (I should like to thank Jim Secord for calling my attention to a cutting of this which he found among Sedgwick's papers in Cambridge University Library.) The *Athenaeum* picked up the story and ran it, which should have left few doubts among the London community (*Athenaeum*, 1414, 2 December 1854, 1463–4). But by then accusing fingers had long been pointing at Chambers. Even the *Athenaeum* admitted that it came as no surprise. "Mr Page fixes the authorship on a gentleman who has been generally credited with the work."

Did Owen know who the author was, count him as a friend, and remain silent out of loyalty? Among Owen's letters is indeed one from Chambers thanking Owen for a copy of *Nature of Limbs*, and mentioning the "pleasure and honour of knowing you personally" (6 March 1849, BM[NH] OC 7.19). So it is conceivable that they were friendly five years earlier, but as to Owen suspecting his authorship there are doubts. Owen clearly *thought* he knew who the author was, since he mentioned being visited by the "reputed Author of the 'Vestiges'" to Chapman in 1848, who replied:

> I presume the "reputed Author of the 'Vestiges'" who paid you a visit was Robert Chambers? I think there is now pretty strong evidence to fix the paternity upon him (Chapman to Owen, 13 January 1848, BM[NH] OC 7.26).

But Owen might not have meant Chambers at all. On the contrary, he seems to have suspected the phrenologist George Combe, family friend and frequent guest of the Chambers household (Gibbon [1878], II, 187–9). Owen detected a "correction of a zoolog.[1] error in the 2[d] Ed." of *Vestiges* (i.e. 1844) which looked like Combe's doing – so he put it to him. Now, Combe had already guessed Chambers' secret and swore never to reveal it, knowing the damage it would do the firm. Therefore, he kept silent when challenged by Owen and unwittingly incriminated himself. Of course, the correction may actually have been Combe's, which Chambers picked up casually in conversation – nonetheless, Owen now had Combe pegged as the author. As a consequence, finding *Vestiges* catalogued under "Chambers" in the Hunterian library ("from gen.[1] rumour & circumstantial evidence" the librarian informed him), Owen had the book removed to the "anonymous" list. (Owen to Sir James Clark, April 1859, BM[NH] OC 7.209.) Of course, the problem remains, did Owen keep silent out of loyalty? But it seems unlikely that his failure to write a damning refutation was out of deference to Combe.

29 Shairp, Tait, & Adams-Reilly (1873), 178.
30 Owen, "Hunterian Lecture, 1837", 30–36, in *Manuscript Notes, and Synopses of Lectures. 1828–1841*, BM(NH) OC 38. Owen (1835) had demolished the Lamarckian threat by reinterpreting chimpanzee anatomy. Bartholomew (1973, 1979) and Ospovat (1977) have suggested that Lyell restructured geology for a similar reason. Brooke (1977b), 142, quotes Owen to Whewell on a baboon great grandparent; and Owen's reply to the Vestigian is in R. Owen (1894), I, 251.

31 Robert E. Grant, "Palaeozoology" Lectures (1853), BL Add. 31,197.
32 Owen to Chapman, n.d., Rev. R. Owen (1894), I, 309–310.
33 Owen (1860b), 495, 496.
34 Owen & Broderip (1853), 68; also Owen's paper "On Metamorphosis and Metagenesis" (1854c), and his book *On Parthenogenesis* (1849c). For a detailed look at 'alternation of generations' consult Mary Winsor's *Starfish, Jellyfish, and the Order of Life* (1976).
35 Owen (1858a), lxxv; Huxley (1894), 324; L. Huxley (1900), I, 94; for Owen's tirade against Darwin's polarised options, see (1860b), 502.

When Huxley's friend Kölliker looked like following suit, Huxley immediately tackled him in print: (1864b), 94–7. Since the cycle returned to its start, Lyell (1863), 421, concluded that, although alternation of generations did enlarge "our views of the range of metamorphosis through which a species may pass", it was ultimately no rival to transmutation.

36 Gillespie (1979), 31; Bartholomew makes the same point as Gillespie: (1973), 288. Rehbock (1978), 341, questions this kind of hindsightful approach and makes some apposite comments, as does W. F. Cannon (1976), 106.

37 W. S. Macleay to Owen, 28 April 1850, BM(NH) OC 18.331. Huxley set out to test Owen's "revived" theory and concluded that the so-called "generations" were no more than "zooids" in (1851a), 38–9. Darwin doubted that he could so easily sweep aside the notion of individuality: Darwin to Huxley, 17 July 1851, IC 5.2. George Allman on the other hand thought Huxley had "triumphantly demolished the whole system of Alternation of Generations and its cousin Parthenogenesis" though adding that he was unfamiliar with all the facts: Allman to Huxley, 30 May 1852, IC 10.63. Huxley's draft of his Friday lecture to the Evening lecture to the Royal Institution on "Animal Individuality" is at Imperial College, see esp. IC 38.7 and 38.42. Cf. Huxley (1852).

38 L. Huxley (1900), I, 80. Huxley's sly attack on Owen in the *Vestiges* review (Huxley [1854], 433, 439) has been commented on by Ruse (1979), 143. Darwin knew this was Huxley's hatchet work, and thought his handling of "a great Professor . . . exquisite and inimitable" (F. Darwin & Seward [1903], I, 75). Chambers must have been in the dark, or else quickly forgiving, since he wrote Huxley a pleasant letter (14 December 1855, IC 12.164) offering to put him up while in Edinburgh.

39 Carpenter to unknown correspondent, 1 September 1855, IC 12.80, undoubtedly concerning Huxley's new review ("Owen & Rymer Jones on Comparative Anatomy") which was indeed written in a spirit of "constant depreciation" and highlighted Owen's "disposition to exalt himself at [others'] expense" (Huxley [1856a], 26, 27). Personal relations had deteriorated dramatically in 1855 after Huxley's review of *Vestiges*: Owen reported one of Huxley's errors (concerning the brachiopod heart), which had undervalued his own work, to William Sharpey, Secretary of the Royal Society (Sharpey to Owen, 12 ?February 1855, BM[NH] OC 23.376). Huxley was forced into a public retraction (Huxley [1854–5], 335) and Owen pressed home in his new edition of *Comparative Anatomy and Physiology of Invertebrate Animals* (1855), 493, with scathing comments about Huxley's "blindness".

Carpenter and George Busk actually interrupted Huxley's honeymoon in Tenby to draw his attention to the slur, advising that "the best proof you can give of your full possession of eyesight, will be to put a bullet into some fleshy part of your antagonist, without doing him mortal damage".

> Busk and I *roared* over his absurdities, which he has the face to put forward as a representation of the state of British Science in 1855. What *will* the Continentals think of this? (Carpenter to Huxley, 16 July 1855, IC 12.78)

And in the *Medico-Chirurgical Review* Huxley did indeed slate Owen's invertebrate anatomy as old hat. One can well understand Huxley's annoyance, since he figured very poorly in what was, after all, a standard reference tome – and one which omitted any reference to his "very pointed refutation" of 'Alternation of Generations'.

40 Huxley to F. D. Dyster, December 1856, IC 15.80. On Owen's *Medical Directory* entry see Huxley to John Churchill, 22 January 1857 IC 12.194 ("Mr Owen holds no appointment whatever in the Govt. School of Mines" and the announcement "is calculated to do me injury".) And Churchill's reply stating that the entry was as furnished on Owen's return (24 January 1857, IC 12.195).

41 Victor Carus to Huxley, 19 March 1857, IC 12.131.

42 Paradis (1978), 31–36, 66–7; Mandelbaum (1971), 15, designates Huxley a "critical positivist".

43 Huxley to F. D. Dyster, December 1856, IC 15.80.

44 Ghiselin (1980), 108; Allen (1978a), 88. On Owen, Dickens, and *Household Words*, see Lohrli (1973), 392–3. Henry Woodward's is perhaps the best and most balanced short biography of Owen, probably because it was written by an 'insider' intimately acquainted with the man. For the upshot of Owen's "love of high and exalted personages", H. Woodward (1893), 52.

45 Carpenter to Huxley, 22 October 1858, IC 12.94. Turner discusses "professionalisation" in two stimulating articles, (1978) and (1980).

46 Huxley (1858), 571, 584–5.

47 Ashton (1980), 126; this is a useful source for understanding the 'Germanising' influence of Carlyle, Lewes, and Eliot. Lewes (1852), 481. Srilekha Bell's (1981) biography of Lewes concentrates on his romantic and biological work, and his slow disenchantment with the archetype. Levere (1981) discusses Coleridge's influence on nineteenth-century science, especially his last chapter which deals with comparative anatomy and John Hunter – and of course Coleridge and Hunter exerted a strong influence on Owen's thought.

48 Huxley to Kölliker, 25 January 1853, IC 19.278.

49 Anon. (1860a), 479; responding to Owen (1860a), 414. F. Gregory (1977) gives a good account of scientific materialism in the Germany of the fifties; see also Temkin (1968).

50 Ashton (1980), 94, 100, 101. Carlyle called Owen "neither a fool nor a humbug": Rev. R. Owen (1894), I, 198. Peckham (1959), 188–9, discusses the "legalists of nature".

51 Owen to his sister Maria, 7 November 1852, RCS, *Richard Owen Correspondence: 1826–1889*, 3.387; cf. Rev. R. Owen (1984), I, 387–8. According to Rudwick (1972), 210–211, Owen had been discussing the "Ideal Archetype" since 1841, and the earliest reference to the word "Homology" I have so far discovered occurs at the end of his "Notebook 12" (BM[NH]), giving it a probable date of late 1837 – i.e. at the height of his anti-Lamarckian stand. Farber (1976), 107–113, looks at the romantic concept of "Type", and Ospovat (1976), 17–24, its application.

52 Owen & Broderip (1853), 50; Owen (1851), 449.

53 Although it did prompt Whewell to ask a teleological question: namely, what purpose did it serve giving the birds, beasts, and fishes, "a skeleton of the same plan, and even of the same parts, bone for bone"? A rather dangerous question, since he could find none, and looked in peril of admitting that this strategy was creatively meaningless (Whewell [1854], 319).

54 Curwen (1940), 232; Chambers to Owen, 6 March 1849, BM(NH) OC 7.19; on Powell, see R. M. Young (1980), 85–6, and on his support for *Vestiges*, the introduction to the twelfth edition, Chambers (1884), xxx–xxxi.

55 Hamilton (1856), cxv–cxviii; the quote from Powell on "our convictions" comes from his *Connexion* (1838), 155–6.

56 The fact that a lawful creation was never taken to imply transmutation explains why Sedgwick could proof-read Babbage's *Ninth Bridgewater Treatise* (1838) – which dispensed with a tinkering Deity and substituted a Divine programmer – and casually dismiss it as "too ambitious" (Clark & Highes [1890], I, 483). Nothing more; there was never any question that it was *dangerous*, or that the Devil was really getting his due (K. Lyell [1881], I, 467). In the same way, Babbage's other proof-reader, William Fitton, was disappointed because miracles as manifestations of some overarching law was "nothing very new", and hardly needed the calculating machine to explain it (Fitton to Babbage, 23 January 1837, BL Add. 37,190, f. 19).

Evidently only Lyell and Powell really appreciated Babbage's intent. The book, Babbage told Princess Victoria, was "written in defence of Science and for the support of Religion" (24 May 1837, BL Add. 37,190, f. 147). He had set out to counter the "Creative Interference" in

Buckland's official Bridgewater treatise, *Geology and Mineralogy* ([1837], I, 586), which he considered an attack on rational science and a weakening of natural religion. It left the Deity incapable of planning ahead, thus "denying to him the possession of that foresight which is the highest attribute of omnipotence" (Babbage [1838], 24–5). He showed with his calculating machine how a new sequence of numbers could be programmed to cut in, just as "at appointed periods" new races of life had made their debut on the earth.

Although the book disappointed its other proof-readers, Lyell thought the chapters on law marvellous, although he questioned some tasteless passages elsewhere in the book (Lyell to Babbage, 17 February 1837, BL Add. 37,190, f. 37; May 1837, f. 185; K. Lyell [1881], II, 9–10). W. F. Cannon (1960), 25, imagined that Babbage's conclusions came as "cold comfort" to the uniformitarians, but this is probably not true. Only part of the correspondence was printed, and a study of Babbage's letters in the British Library suggests that Lyell was distinctly impressed.

57 Curwen (1940), 261.

58 Powell to Owen, 25 January 1850, BL Add. 42,577, f. 40. Owen relegated final causes to a subordinate role in his *Nature of Limbs* (1849a), a point naturally endorsed by Darwin in the *Origin* (1859), 434, 435. Dov Ospovat (1978), 49–52 suggests we shift away from the old 'Creation vs Evolution' dichotomy and use individual responses to final causes to achieve a less distorted view of contemporary alignments.

59 Owen to Powell, 26 January 1850, in the possession of Mr Donald Baden-Powell (I should like to thank Jon Hodge for showing me a copy); cf. Brooke's commentary on this letter: (1979), 40–41. Also Brooke (1977b), 143, republishes Owen's reply to the *Manchester Spectator* entire.

60 Powell (1856), 102, 382–7, 392–3, 408–9, 417–8, 477–8; endorsing Owen (1849), 86.

61 Owen to Babbage, 23 July 1856, BL Add. 37,197, f. 73; Hamilton (1856), cxv–cxviii.

62 Owen (1859a), 62; cf. Guterman's introduction to Schelling (1966) and Gray-Smith (1933), 45.

63 Owen to Rudolph Wagner, 13 February 1849, RCS Stone Collection, 4.

64 Darwin (1859), 416; Russell (1916), 247, long ago realised that the morphology of the fifties "could be taken over, lock, stock and barrel, to the evolutionary camp". On Huxley's resistance to Owenian specialisation, see Bartholomew (1975) and my Chapter 3.

65 Blake (1863), 153. Huxley (1854), 425, 426, 427, 429.

66 Huxley (1853), 176–7, 192; for more idealised conceptions, cf. Owen & Broderip (1853), 70–83. Lewes (1852), 498–9, is discussed by Bell (1981), 288–290. On Huxley's lingering romanticism, see Paradis (1978).

67 Darwin to Huxley, 23 April 1853, IC 5.4; cf. F. Darwin & Seward (1903), I, 73.

68 Ospovat (1980), 175, (1981), 72–3, 146. Contrast his views on Darwin as a "theist" with H. E. Gruber (1974), 208–213, 314, and Manier (1978), 204 n. 13. Greene (1975), 246–7, 250, also believes Darwin was an "evolutionary deist" at the time he wrote the *Origin*, a subject studied in some detail by Gillespie (1979).

69 Huxley to J. D. Hooker, 18 June 1858, IC 2.153.

70 T. J. Parker (1893), 125, also 44, 66, 79, 108, 113; L. Huxley (1900), I, 271–2, 415.

71 W. K. Parker (1864b), 341.

72 Cornish (1904), 92–3; on Baden Powell, 43, 45–7 (Powell had married the Flowers in 1858 and two years later the young anatomist kept a three-day vigil at Powell's deathbed). On the "ecclesiastical swallow", L. Huxley (1900), II, 56.

73 Flower (1898), 133, 134. Lydekker (1906). 172.

74 Flower (1870), 196, 199.

75 Rolleston to Huxley, 20 May 1860, IC 25.147; and pledges his support in n.d. IC 25.148; while persuading his friends that the Established Creed can be reconciled with Darwin in, 13 April 1860, IC 25.142. Tylor's biography of Rolleston is the most detailed (Rolleston [1884], I, ix–lxxvi). Huxley discussed him in L. Huxley (1900), I, 191, and the *Review*, 209–210. Rolleston (1861), 56, 61, denounces "Platonic mysticism" and talks of the failing empire of the archetype.

76 L. Huxley (1900), I, 249 1250, 270; T. J. Parker (1893), 84

77 Cornish (1904), 100; Flower to Huxley, 11 July 1862, IC 16.117; L. Huxley (1900), I, 235. Huxley and Flower collaborated during the sixties in the "green oasis" of Lincoln's Inn Fields, and many an anti-Owenian plot was hatched either there or in Jermyn Street. When pressure of work forced Huxley to resign, Flower was his obvious choice as successor (L. Huxley [1900], I, 312, 249–250; Cornish [1904], 102–3). Flower's inaugural on taking the Hunterian Chair in 1870 was defiantly anti-Platonic, despite Owen's presence in the audience. At the time he applied for the Conservatorship in 1861, Parker lived off his medical practice in Bessborough Street. He was not unduly dismayed by Huxley's support for Flower (his junior by eight years), and rather touchingly told Flower "If your big *farm* lay close to my little *garden* I should often put my bony leg over the fence and be your voluntary assistant from mere love of the place and the work" (T. J. Parker

[1893], 31–2; Cornish [1904], 51). And as if to consolidate the anti-Platonic feeling at the College of Surgeons, Parker did eventually climb the fence, joining Flower in the second Chair of Anatomy and Physiology in 1873.

2. Creative Continuity

1 Owen's criticism of the *Origin* in the *Edinburgh Review* (1860b), 511, was republished by Hull (1973). Huxley calls Owen an advocate of "received doctrines" in (1860a), 28.

2 Eiseley (1979), 5, 41. Rather than throwing any "New Light", this book was put together posthumously from articles published between 1956 and 1972, and much of the material has been superseded. Moore (1979) discusses the "military metaphor".

3 Huxley (1893a), 15; L. Huxley (1900), I, 109

4 Huxley to F. D. Dyster, 3 November 1856, IC 15.78; for Huxley's attack on Cuvierian methods, see (1856b), and Darwin's consideration of this F. Darwin & Seward (1903), I, 89.

5 Huxley (1854), 432; Clark & Hughes (1890), I, 447. Agassiz (1843), 16, used the ice-sheet "like a sharp sword"; and as late as 1865 he set out for the Amazon basin to prove that it was once glaciated and that ice had shrouded the entire planet, his intention plainly being to refute Darwin's views: L. & E. Agassiz (1868), 15, 33, 399, 425; see also F. Darwin & Seward (1903), II, 159–161, and K. Lyell (1881), II, 410.

6 Forbes (1853), xxviii–lix; de Beaumont (1852); F. Darwin & Seward (1903), II, 159. For the geological community's response to the rival systems of Lyell and de Beaumont in the 1830s, see Lawrence (1978).

7 C. Lyell to R. Owen, 19 October 1846, BM(NH) OC 18.136.

8 Huxley (1880a), 397.

9 C. Darwin to T. H. Huxley, 1 January 1860, IC 5.94; cf. F. Darwin & Seward (1903), I, 135. Huxley's *Times* review is (1859b).

10 C. Darwin to T. H. Huxley, 28 December 1859, IC 5.92, which is the postscript to F. Darwin (1887), II, 253.

11 F. Darwin & Seward (1903), I, 149; F. Darwin (1887), II, 240, 312; Ospovat (1980).

12 St G. Mivart to Owen, 13 April 1872, BM(NH) OC 19.270; J. Gruber (1960), 20; Mivart discussed the "invigorated" Philosophical Anatomy in (1875), 278.

13 Anon. (1860a), 479.

14 Owen (1858a), li.

15 Stearn (1981), 42–3; Rev. R. Owen (1894), II, 73.

16 Politically, Owen was a Gladstonian Liberal who was to become

Vice-President of the Richmond Liberal Association (Owen to J. K. Langdon Edis, 5 letters, 1882–5, BM[NH] OC). He was also a personal friend of Gladstone's, and as late as 1885 was proof-reading his papers, when the Opposition leader was doing pious battle with Huxley in *The Nineteenth Century* over the compatibility of Genesis and geology. (Helfand [1977] puts the political background to this debate; see also Moore [1979], 65–7, and Jensen [1970] on the *x* Club's dislike of Gladstone's Irish policy and wish to see him discredited; MacLeod [1974], 52–3, 61–2, tackles Owen's alliance with Gladstone against the *x* Clubbers at the time of the Ayrton incident.) Being accustomed to imagine Owen "coaching" Wilberforce, we expect him here to be striking another vicarious blow at Huxley and Darwin. Fortunately Owen's letters to Gladstone are preserved and tell the true story. Owen was making the same points privately that Huxley was to make in print, educating the Liberal leader on fossil continuity and pointing out facts not in accord with Genesis. Owen had misgivings about mixing science and Scripture. "The divine Chapter teaching the fundamental Truth of the guidance to a Higher Life", he advised Gladstone, had simply used "phenomena as were intelligible to the age it addressed". Genesis was a moral guide for the uneducated and as such useful "to the non-scientific and the wage-classes of all time". Its role was assured because the "instances of creative Power" revealed by palaeontology "are unintelligible to many classes requiring to be taught His Will." But it was not a scientific textbook, and should not be set alongside accurate geological determinations of fossil succession. Although Owen tried to pull Gladstone into the present, there was no glorying, Huxley-style, in a good episcopal pounding, and yet he could agree with him on the interpretation of the fossil record and the palaeontological irrelevance of Genesis. (Owen to W. E. Gladstone, 5 January 1884, 7 and 14 December 1885, BL Add. 44,485, f. 32; 44, 493, ff. 188, 214, 223.)

17 T. H. Huxley to W. Sharpey, 13 November 1862, UCL Sharpey Correspondence MS. Add 277 (no. 122).

18 Owen (1860a), 403; (1858a), xlix–li; his Fullerian lecture "On the Extinction of Species" (1859a), 62–3, was published as Appendix A to *On the Classification and Geographical Distribution of the Mammalia*. Hull (1973), Chapter 3, goes more deeply into contemporary philosophy of science. Argyll congratulated Owen on his distinction between Creation by Law and the hypothetical mechanism in a letter: 27 February 1863, BM(NH) OC 1.230.

19 Huxley (1863a), 106; Owen responds to Flower's charge in (1874–89), 87, although I have not seen Flower's original quote. Owen's convoluted "*continuous operation . . .*" is in (1860b), 500.

20 On Darwin's use of Pentateuchal terms (for which, ironically, he was criticised by Owen), see C. Darwin (1859), 488, 490; F. Darwin (1887), III, 18 and II, 251. W. F. Cannon (1961), 131–2, provides a modern discussion which contrasts in approach with Gillespie's *Charles Darwin and the Problem of Creation* (1979). As always, Ospovat (1978, 1980) shows sympathy and understanding. H. E. Gruber (1974), Chapter 2, "The Threat of Persecution" and *passim*, deals most effectively with Darwin's fears and his strategy for disguising his materialism.

21 Powell (1859), 230. Bowler (1977a) most recently has called for a reassessment of this mode of arguing for design in nature. Brooke (1979), 41, 56 n. 14, rightly points out that Owen's references to a continuity of secondary causes, often confused with a belief in 'evolution' (MacLeod, 1965, 261; Bowler, 1974, 183 n. 48), plainly had a theological as much as scientific import. He was combining a Powellian natural theology with a scientific rejection of transmutation.

22 Meyer (1848), 52.

23 Owen (1859b), 154.

24 Owen (1841), 197; Desmond (1979).

25 Huxley (1868a), 304. Compare Huxley's lecture on "intermediate" animals with Owen's "Report" on fossil reptiles (1859b), drawn up nine years earlier, and in which he was clearly parading "annectant" forms. Apparently only Gaudry (1862–7) took any notice of Owen's action.

26 Meyer (1848), 55; Goldfuss (1847), 3–12. On Huxley's and Owen's contrasting restorations of labyrinthodonts, see Owen (1854a), 36, and Huxley (1859a), 270.

27 The Earl of Enniskillen to R. Owen, 19 August (no year), BM(NH) OC 11.229. Illustrations of the Goldfuss specimens of which Owen had casts can be seen in Goldfuss (1847), Plates I–III.

28 Owen (1854b), 117–9. His interleaved copy of the *Catalogue* is housed in the Hunterian Museum of the Royal College of Surgeons (No. RCS 3.1).

29 Portlock (1858), xxi-xxii, made the presentation. Meyer's new work on *Archegosaurus* was his (1856–8), 59–220, and for Portlock's appreciation of this see (1858), ciii-civ.

30 Owen (1859b), 154–6, is the main source of information on his attempt to bridge the classes. I have seen the printed syllabus but not the manuscript notes (if extant) of his lectures at the School of Mines: "Synopsis of a course of Lectures on Fossil Birds and Reptiles . . ." delivered in the Museum of Practical Geology, Jermyn Street, lectures 3 and 4, 25 and 26 March, 1858: *Richard Owen: Manuscripts, Notes, and Synopses of Lectures*, Vol. 3 (1849–64), BM(NH).

31 Bowler (1976), 106, 110. On Miller's nature "straight as an arrow", see

(1857), 112. This was Miller's last book, and he was shortly to shoot himself in a visionary fit. Ospovat (1981), Chapter 5, has shown that the branching conception of nature as a deduction of von Baerian science was perhaps more common in pre-Darwinian science than was previously supposed.

32 Robert E. Grant, "Palaeozoology" Lectures (1853), BL Add. 31,197, esp. ff. 17, 18, 20.

33 Owen (1859c), 128. Powell in his *Essays* (1856), 2nd ed., 439, reports that Owen evoked the image of a fossil "net-work" in a lecture to the British Association at Liverpool in 1854, although I have not been able to trace it. However Owen does use the word "network" in his review of the *Origin* (1860b), 493.

One wonders to what extent the analogy of extraterrestrial life influenced Owen. He certainly read Whewell's *Plurality of Worlds* (1854) – which argued against extraterrestrial civilization – "with a sense of dissatisfaction, and of pain" and as an example of "special pleading" unworthy of the man (Owen to B. Dockray, 1 April 1854, RCS, *Correspondence of Richard Owen, 1826–1889*, 3.438–9; cf. BL Add. 33,348 ff. 12, 39). It presumably made no sense theologically to depopulate the heavens, leaving them empty and wasted, and Owen may have favoured "mutation" or diversification on the earth, and an exploitation of the archetype's full potential, for similar aesthetic reasons (all of which rested, as Buckland long ago affirmed, on the belief that life was "created first for its own sake" – [1837] I, 99n, 101).

34 Rudwick (1978), 95–6.

35 Huxley (1863b), 66–8, replying to Owen's (1858b), 155. On the creation of the order Dinosauria see Desmond (1979). On the subsequent uses of the Ganocephala see Haeckel (1876), II, 216, and Cope (1887), 332. Osborn (1931), 243, 247, 253, speaks of Cope's friendship with Owen; Huxley, on the other hand, seems to have been indifferent to Cope.

36 Portlock (1858), lxxxii. For Murchison's appreciation of Portlock – "a geologist quite after my own heart" – see Murchison (1864), cxv–cxvii.

37 J. Wyman to R. Owen, June 1863, BM(NH) OC 27.254; Argyll (1868), 219 (and on Argyll see Gillespie, 1979, 93–104). The *Quarterly* reviewer was J. B. Mozley (1869), 162. On Owen and the Darwinians generally see MacLeod (1965), 263–4, 272–8; Hull (1973), 171–5, 213–5; Ruse (1979), 228, 242–3, 260–1. Owen's own *Edinburgh Review* critique (1860b) repays careful study, as does his holograph article "Species" (c. 1860) among the *Autograph MSS of Sir R. Owen*, BM(NH) OC 59.13. To understand the origin of Owen's physiological approach to the species question, it is important to consider his third

Hunterian lecture for 1837: *Manuscript Notes, and Synopses of Lectures. 1828–1841*, BM(NH) OC 38, lecture 3, esp. pp. 30–6.

38 G. Rolleston to T. H. Huxley, 13 April 1860, IC 25.142. For Darwin's comments on Owen's severe and damaging review, F. Darwin (1887), II, 300–1, and on Owen's envy, F. Darwin & Seward (1903), I, 149. Owen's accusation of "indifference" (1860b), 500–1, and his cry of "preposterous" (1866–8), III, 796, n. 2, were responses to Darwin's caricature, C. Darwin (1859), 482–3. For *The Saturday Review* riposte see Anon. (1860b) and F. Darwin (1887), II, 311. Hull (1973), 408, speculates on Mivart's authorship of this piece. Darwin's breaking off communications with Owen from this time on had predictable consequences; when, for example, he was sent a fossil horse molar from America and asked to forward it to Owen, he sent it to Huxley instead for study; Darwin to Huxley, 21 September 1869, IC 5.273.

39 Owen (1863c), 62 n. 1, being a response, oddly enough, to Grant – and one which must have amused the old heretic immensely. Owen again made the point about human "mutation" in (1866–8), I, xxxvi. The anti-Lamarckian slant of Owen's earlier ape work can be found in Owen (1835), 370–2, and (1849b), 414–7. On Savage's discovery of the gorilla see T. Savage to R. Owen, 24 April 1847, BM(NH) OC 23.103. Savage had been sending Owen chimpanzees since 1843, OC 23.123. For a parallel analysis of Lyell's response to the Lamarckian threat, Bartholomew (1973, 1979), Ospovat (1977). On Lyell's attempt to rationalise mankind's "birth", see Wilson (1970), 86, 119, 127–8, 153, 243, 261, 279, and C. Lyell to T. H. Huxley, 17 June 1859, IC 6.20 (cf. Wilson [1970], 261). The syllabuses of Owen's Hunterian lectures show that whenever he treated ape anatomy, he always ended with a refutation of the "transmutation-hypothesis".

40 W. Whewell to R. Owen, 3 April 1859, BM(NH) OC 26.285. Owen's lecture "On the Gorilla" was printed as an appendix (B) to his *Classification* (1859a). On man's adaptation to become the seat of a "responsible soul", Owen (1835), 343; and the subclass Archencephala was announced in Owen (1858b), 19–20. Argyll (1869), 62–3, supported Owen on the vexed question of man's taxonomic distinction.

41 W. Vrolik to R. Owen, 14 November 1849, BM(NH) OC 25.353. Lankester (1899), 254, discusses Owen's errors. Oddly, Owen had been dealing with the brains of apes and man and the problems of spirit preservation since about 1830: cf. his MS. Notebook 1 (Oct.–Dec. 1830), BM(NH) L. OC.o.o.25, for 27 October 1830.

42 T. H. Huxley, "The Principles of Biology", 14 Royal Institution lectures delivered 19 January–23 March 1858. Lecture 10, "On the Special Peculiarities of Man", 16 March 1858, IC 36.96–108, esp.

36.97, 36.99. Huxley (1863a), 118, boiled down the issue to one of "personal veracity".

43 G. Rolleston to unknown correspondent, 1 October 1861, Wellcome Institute for the History of Medicine, London, AL 325619; Rolleston (1884), I, 21, 52. Darwin clapped Huxley on the back in C. Darwin to T. H. Huxley, 1 April 1861, IC 5.162. Owen's "Ape-Origin of Man" title is his (1863b), and he accused Huxley of advocating "man's origin from a transmuted ape" in (1861b), 395.

44 T. H. Huxley to W. Sharpey, 13 November 1862, UCL Sharpey Correspondence MS. Add 227, No. 122. Sharpey (1862), considers Owen's archetypes of "surpassing beauty" – see also Taylor's biography (1971), 153, 242. Sharpey's reply ("No one is more opposed to him . . .) is 13 November 1862, UCL Sharpey Correspondence MS. Add 227 (No. 123); Huxley's rejoinder, 16 November 1862, ibid. No. 124. And Edward Sabine's letter to Sharpey on this moral dilemma, 14 November 1862, ibid. No. 121.

45 T. H. Huxley to F. D. Dyster, 11 October 1862, IC 15.123.

46 C. Lyell to T. H. Huxley, 5 July 1861 (incorrectly dated 1860), IC 6.36. T. H. Huxley to F. D. Dyster on the "second thousand", 12 March 1863, ibid. 15.125. Despite Wagner's initial letter to Owen (30 November 1860, BM[NH] OC 26.12), his paper (1861) and subsequent letter to Huxley (4 January 1863, IC 28.88), show him reluctantly siding with the 'opposition' on *facts*, while praising Owen's idealistic intent. W. C. Thomson's letter to Huxley on his fight with Halford (26 November 1863, IC 27.332) shows how quickly contemporary issues were taken up in the new colonial universities – on which see A. Mozley (1967), 425; Moyal (1975).

47 Kölliker thought "alternation of generations" (what he called "heterogenesis") could give some insight into the production of new species, and presumably Huxley (1864b), 94–7, singled his analogy out for attack because he knew it was also Owen's line of reasoning. Ordinarily Huxley had a great respect for Kölliker, even though he was an idealist who more closely approached Mivart in his world view.

48 G. Rorison to R. Owen, 25 April 1860, BM(NH) OC 22.379, which also contains Owen's reply.

49 Rorison (1862), 322.

50 Owen (1864), 27, 31. *The British Standard*, 5 February 1864 (and a reply on 12 February).

51 C. Lyell to T. H. Huxley, 20 February 1864, IC 6.97. The *Geological Magazine* review was Anon. (1864), 82. C. C. Blake wrote to Owen after the lecture, 22 December 1863, BM(NH) OC 3.204 (it is conceivable of course that the printed lecture was stronger in tone than the oral one – on which, see Gunther [1980], 125). Henry Lawson thanked

Owen for the "glorious" manner in which he had "overturned the absurd notions of certain evangelical folk", and requested a follow-up paper on the same subject for the next number of his *Popular Science Review*: H. Lawson to R. Owen, 4 July 1864, BM(NH) OC 17.219. Huxley resigned from the Anthropological Society ostensibly because of Blake's "coarse attack" on Rolleston in his article "Man and Beast" (Blake, 1863), but since the article in question was a general condemnation of *Man's Place* Huxley would have had a more personal reason for resigning: T. H. Huxley to C. C. Blake, 2 May 1863, IC 11.17. Huxley called them all "quacks" in a letter to J. Fayrer, L. Huxley (1900), I, 274.

52 C. Lyell to T. H. Huxley, 9 August 1862, IC 6.66. In the same vein Lyell complained about the tone of one of Carpenter's Sunday Evening Lectures, in which he had attacked Calvinism (6.122). Huxley particularly upbraids "Parsondom" in his letters to F. D. Dyster, and quoted are those of 30 January 1859, ibid. 15.106 and 29 February 1860, ibid. 15.110. Moore (1979) provides the best critique of Huxley's position.

53 *The Daily Telegraph*, 10 April 1863, p. 4. MacLeod (1965), 273 n. 66, deals with Owen and Wilberforce. Of course, "round-mouthed, oily, special pleading" is Huxley's description in a letter to Dyster, 9 September 1860, IC 15.115. For a reappraisal of the "legendary encounter" between Huxley and Wilberforce, see Lucas (1979).

54 Anon. (1860a), 478–9.

55 Owen (1861a), 204–5; cf. the 1st edition of *Palaeontology* (1860a), 182.

3. Huxley's 'Persistence'

1 Huxley (1869a), 390. The extreme antiquity of mammals was 'deduced' from their state of development in Mesozoic times: Owen (1857) himself insisted that the Jurassic *Stereognathus* was already recognisable as a pig-relative, and Huxley was equally adamant that Mesozoic 'marsupials' were highly developed, suggesting a long prior history. Bowler (1976), 75, makes this point. Duffin (1978), 60, discusses the Triassic mammals collected in Britain at this time. For a rival opinion that birds and mammals had a recent (Triassic) origin, see Milne Edwards in *Geol. Mag.*, 1 (1864), 71.

2 L. Huxley (1900), I, 173. Poulton (1908a) early appreciated Huxley's divergent views; Bartholomew (1975) has recently looked at his "defence of Darwin", and Winsor (1976) the attraction of Macleay's circular system. Dov Ospovat, in his unpublished "Darwin and Huxley on Divergence: Some Darwin Notes on His Meeting with

Huxley, Hooker, and Wollaston in April, 1856", also devoted time to Huxley's dislike of progressive divergence and his flirtation with the circular system. Greene (1975), 249, commented that no attempt has been made to distinguish Darwin's Darwinism from Huxley's.

3 Huxley (1863a), 159; L. Huxley (1900), I, 174 – for Lyell's original letter to Huxley on 17 June 1859, see IC 6.20, and the extract in Wilson (1970), 261–2, where he wonders about finding "fossil men intermediate between some extinct Quadrumana & the Bimana".

4 Huxley (1863b), 565; see also Huxley (1862c) and L. Huxley (1900), I, 263–4.

5 Huxley (1870a), 528.

6 F. Darwin & Seward (1903), II, 234, replying to L. Huxley (1900), I, 205–6.

7 It was, however, true, as Marsh said in 1877, that *Lepidosiren*, "although its immediate predecessors are unknown, has some peculiar characters which strongly point to a Devonian ancestry" (1877), 8, so its use by Owen was quite acceptable.

8 Huxley (1869b), 41.

9 Only in the later 1860s and early 1870s did Huxley incline towards Owen's explanation, i.e. that amphibia had a *Lepidosiren*-like ancestry. For example, in reviewing Haeckel's *History of Creation*, Huxley suggested that "the Crossopterygian Ganoids exhibit the closest connection with *Lepidosiren*, and thereby with the *Amphibia*", (1869b), 42, and he went on to trace a line from the Cyclostomes or jawless lampreys and hagfishes through the Ganoids and "Mudfishes" (lungfishes) to the amphibia. Cf. Huxley (1870a), 549. The fact that Huxley's strongest statement was made in a Haeckelian context reinforces my thesis in Chapter 5 that Haeckel was perhaps the main instigator at this time of Huxley's change of direction.

10 L. Huxley (1900), I, 149; Huxley (1862a), 525–6; Lovejoy (1968), 391–4. In the same way, in his Royal Institution lecture "On Species and Races, and their Origin" (1860b), 389 – the lecture Darwin considered such "an entire failure" – Huxley looked straight at Owen and flippantly insisted that *Palaeotherium* was not more "generalized" compared to later horses, any more than a father is to his sons. Owen was amazed, and of course the whole thing rebounded in Darwin's face. "By their fruits may the promoters of true and false philosophy be known", Owen seethed in his "atrociously severe" review of the *Origin*, (1860b), 521; F. Darwin & Seward (1903), I, 139; Bartholomew (1975), 531; Hull (1973), 171–2.

11 K. Lyell (1881), II, 356.

12 Wilson (1970), 185; Lyell (1851) and (1852), "Postscript" inserted at the beginning of the *Manual*, vii–xxii, on which see Bartholomew

(1976), 168–9. Incidentally, by 1859 Lyell had accepted Huxley's view that *Telerpeton* was late Triassic: Huxley to Lyell, 10 October 1859, IC 30.33.

13 Actually, Owen snatched the *Dryopithecus* from Lyell's grasp, although it did originally look like a candidate for human ancestry. The jaws and upper arm bone of this huge Miocene ape had been exhumed by Edouard Lartet (1856) in Saint-Gaudens. Because it was man-sized and its milk tooth replacement was more like a man or gibbon than any other ape, Lartet placed it one step nearer humanity than the gorilla. Hence Lyell declared that in stature and structure, it "came nearer to Man" than any ape: (1859), 14–15. So "near the negro" in fact, that "tho' extinct perhaps for a million years & more", it might still fool a student sitting his "osteological examination" in the College of Surgeons: Wilson (1970), 157. As Lyell jotted in his *Journal*:

> If Man was develop.[d] out of any ape it was from the Dryopithecus & a million or more of years have been required for it.

So *Dryopithecus'* strategic importance *viewed evolutionarily* was already frighteningly apparent, and Owen evidently realised the explosive nature of Lyell's position. Lyell sent Owen a copy of his *Supplement* (1859) containing his more guarded views, and Owen promptly pencilled the offending passages relating man anatomically to the fossil ape (see Owen's copy of Lyell's *Supplement* in the British Museum [Natural History] Palaeontology Library). He insisted that the teeth were a gibbon's, and that Lyell's rash statement that the fossil closely approached man "is without the support of any adequate fact", while in reply to Lyell's expectation of finding a still closer fossil form, we must patiently wait until the "world has furnished us with the proofs that [such] a species did formerly exist . . .' Owen (1859a), 86–7.

14 Wilson (1970), 86, 153. Bartholomew (1973) discusses the meaning of these *Journals*, see esp. 290–303; also his papers (1976) and (1979), and Ospovat's (1977) supporting study.

15 Lyell to Huxley, 3 May 1862, IC 6.60, responding to Huxley's anniversary address (1862a), 528. Huxley cautioned Lyell that Marsh had by no means proved that his vertebrae were ichthyosaurian, and they were equally likely to have been labyrinthodont, 5 May 1862, IC 30.40, although cf. Lyell (1863), 403. Lyell discussed von Meyer's *Archegosaurus* in Wilson (1970) 240, 242, 331. Despite the collapse of Marsh's counter example, within a few years Huxley had sniffed out "a quantity of Carboniferous corpses" all labyrinthodont, and all with "ossified spinal columns", which tended to destroy his originally-mooted progressive sequence, L. Huxley (1900), I, 263–4.

16 K. Lyell (1881), 268; Bartholomew (1975), 527; for recent discussions

on Lyell's strategy, see Rudwick (1970), W. F. Cannon (1976), and Porter (1976).

17 F. Darwin (1887), III, 10; Wilson (1970), 118, 327–8. One of Huxley's strongest denials of any "palaeontological evidence" of "progressive change" is in (1865b), 175, which was actually written before Darwin published the *Origin*. Although in this section I have concentrated on Huxley's anti-Owenian and anti-realist strategy, and (following Bartholomew) his support of Lyell's naturalism, Ospovat in his MS "Darwin on Huxley and Divergence" (note 2 above) speculates that Huxley actually misinterpreted von Baer to believe that all vertebrates departed *equally* from the archetype, and this, together with his addiction of Macleay's circular system, confirmed his belief that no progress had occurred.

18 Carpenter to Owen, 23 September 1842, BM(NH) OC 6.308, and 4 July 1845, ibid. 6.329. On Carpenter's friendship with Huxley, see L. Huxley (1900), I, 92, 96, 129; and on his belief that Owen was 'confused' over "Parthenogenesis", Carpenter (1884), 141–2. On his priority dispute with Owen over who first applied von Baer to the fossil record, Carpenter to Owen, 22 October 1851, RCS, *Richard Owen Correspondence, 1826–1889*, 3.366, and cf. BM(NH) OC 6.335–8. For a study of this whole subject see Ospovat (1976).

19 Carpenter (1851), 569, 576–580, and (1854), 110; Ospovat (1981), 117–124, discusses von Baer's embryology and its impact in far greater detail.

20 Huxley (1855), 241–7. On Carpenter's belief that the "application of Von Baer's great doctrine of *Development from the General to the Special*" paved the way for the *Origin*, (1888), 107. Ospovat, "Darwin on Huxley and Divergence", discusses Darwin's assessment of Huxley's criticisms. Stauffer (1975) had edited and published the second part of Darwin's "big book" *Natural Selection*.

21 Huxley (1859c), 93. Lyell recorded in his *Species Journals* on 20 May 1859 that "Huxley believes in it [progression] but suppose[s] that most of the progress was made in times anterior to our oldest fossilif.ⁱ rocks", Wilson (1970), 240.

22 Falconer (1863), 251–4; P. M. Duncan (1865), 361. Huxley's colleague Robert Etheridge, the Palaeontologist to the Geological Survey, mentioned the 'persistence' of armoured dinosaurs through Jurassic and Cretaceous times, (1867), 69.

23 Spencer (1857), 148–9. Gould (1977), 114, comments on von Baer's own cosmic ambitions.

24 Peel (1971), 97; Hofstadter (1955), 33, deals with Youmans; Wiltshire (1978), 73, talks of Spencer's "self-enslavement". Peel's biography is excellent on Spencer's Dissenting roots. Wiltshire, who is strong on

his political growth, concludes that "Spencer's evolution was allowed to develop out of his already finalized political convictions . . ." (66). For the complex interplay between politics, biology, and social theory, see Greta Jones (1980).

25 Peel (1971), 132; as R. M. Young (1965) said, it was "Spencer's belief in rapid social progress which most strongly influenced the way he viewed the biological evidence".

26 D. Duncan (1908), 87; L. Huxley (1900), I, 161; Spencer (1858), 400; Spencer (1904), I, 368 and 462, and II, 24; Peel (1971), 29.

27 D. Duncan (1908), 61; Spencer (1904), I, 384; C. U. M. Smith (1982), 58. See also Spencer's "Illogical Geology" (1859).

28 Spencer (1858), 415–6; Kennedy (1978), 72. Chapter 6 of Professor Kennedy's biography is particularly good on the Spencer–Owen connection.

29 Spencer (1904), I, 408 and II, 476; L. Huxley (1900), I, 333; D. Duncan (1908), 81. Greene (1981), 141–2, underestimates Huxley's geological doctrine in suggesting that he shared Spencer's faith in "Nature's great progression". Or rather, this was truer of Huxley in the 1870s than the 1850s.

30 Spencer (1857), 450; D. Duncan (1908), 83.

31 D. Duncan (1908), 91, discussing Huxley (1859d). On the debate over the age of the Elgin sandstones, see Benton (1982). In "Illogical Geology" (1859), 367–376, Spencer was working on a model of progressive migration onto newly-elevated land which could *simulate* progressive development, even though all development might have taken place in Huxley's "Azoic" times.

32 Spencer (1864), I, 323–6, and (1862), 153–5. The block quote "While we are not called on . . ." comes from Spencer (1864), I, 430–1.

33 With the New World philosophy of Fiske, Sumner and Spencer pre-eminently one of social and industrial progress, Huxley's problem of fossil stagnation looked completely out of place. The "beneficent necessity" of progress was proclaimed throughout the gilded age – the "current of life", Clarence King (1887), 469–470, declared after the revelation of Marsh's horses, must sweep "onward and upward to ever higher and better manifestations". Hence while Huxley met with a rapturous American welcome in 1876, Spencerians, Neo-Lamarckians, and Cosmic Theists were careful to take from him only proof of fossil progress. E.g. Fiske (1939), 121, declared himself "quite wild over Huxley", and following the *American Addresses* he too played horses as his trump card (1876), 28–30. Yet he simply ignored Huxley's "indifferent" evidence, i.e. "persistent types".

The same might be said of palaeontologists of the 'American school'. Cope's original *Vestiges*-like theory (1868–71), in which an

environmental drive ensured foetal "acceleration" and continual progress (Bowler, 1977; Gould, 1977, 85–96) left him unsympathetic to
Huxley's 'persistence', and he argued for the "successional complication of structure among Vertebrata in time": Cope (1870), 137.
Admittedly he was later to investigate "retrogressive evolution, but
this was an *active* process, e.g. Cope (1885), 335, talked of "creation by
degeneration" in the case of the snake – see note 51 below.

Even Darwinians were well grounded in Spencer. Marsh (1877), 48,
accepted natural selection "in the broad sense in which that term is now
used by American evolutionists", and helped get Spencer's banned
Study of Sociology into the classrooms at Yale. At the Peabody Museum
in 1882 Marsh entertained Spencer and at the farewell banquet for the
philosopher at Delmonico's in New York flatteringly echoed
Spencer's words by talking of evolution as the "law of all progress":
Marsh (1882), 3. Try as men may, Fiske was to say, "there is no nook
or corner in speculative science where they can get away from the
sweep of Mr Spencer's thought", (1890), 47. Accordingly, Marsh
accepted the Owen–Spencer "Law of Progress from the General to the
Special" (Owen, 1871, 94, 111, 114 – his *Monograph* on Mesozoic
mammals, which Marsh still considered the "main authority" in 1887,
328). Marsh plotted the progressive expansion of the brains of mammals, birds, and dinosaurs (1879), 42, and of course did something
very unHuxleyan – he created the concept of a "generalized" order of
Mesozoic mammals: see my Chapter 6.

34 Thomas (1888), 311–2, whose conclusions were repeated by Flower &
    Lydekker (1891), 141; Murray (1866), 283. Owen (1857) on *Stereog-
    nathus* provides one of the most prominent exceptions to the prevailing
    belief that Mesozoic mammals were almost all marsupial.

35 Unger (1865b), 45.

36 McCoy (1862), 145, also 137–8; and on his life see H. Woodward
    (1905). For an overview of the Australian situation consult A. Mozley
    (1967), Moyal (1975), and Oldroyd (1980a) for a more general view of
    some Darwinian "impacts". On the colonial image in biogeography,
    see Nelson (1978), 299.

37 Huxley (1869a), 388.

38 Huxley (1870a), 547.

39 L. Huxley (1918), I, 445; Murray (1866), 34; Unger (1865a). For a
    critique of Heer's and Unger's views, and a restatement of faith in
    migration across the Bering Strait, see Oliver (1862) in Huxley's
    Natural History Review.

40 Wallace (1876), I, 156; Fichman (1977); F. Darwin & Seward (1903), I,
    486; F. Darwin (1887), I, 485, and II, 147 where Darwin leans "to the
    side that the continents have since Cambrian times occupied their

present positions." Darwin made his position plain in the *Origin* (1859), 358, as he had done in *Natural Selection*, Stauffer (1975), 538, note 3. Murray (1866), 13, would have nothing to do with an "infinity of experiments"; and on his "hostile review" and Darwin's post-*Origin* impression of him, see F. Darwin (1887), II, 261–2, and III, 230; and F. Darwin & Seward (1903), II, 3, 7, 12, 30, and I, 176, where he groups Murray with those who "sneer" at him. Ghiselin (1969), 39–40 argues that Darwin's non-"extensionist" biogeography was consistent with his general evolutionary methodology.

41 Marsh (1877), 43, and (1878), 459. Actually two jaws of a supposed Triassic marsupial were already known from North Carolina, even if they later turned out to be wrongly diagnosed, Schuchert & LeVene (1940), 446–7. Fichman (1977), 55.

42 P. M. Duncan (1877), 87; Montgomery (1974), 98–101, gives a good account of the German idealist response to Darwin; and J. W. Gruber (1960), esp. 33, is still the standard source for Mivart.

43 On Huxley as "hero", L. Huxley (1900), II, 423; "stormy petrel" in Bibby's expression, and he also quotes Lankester on his "father-in-science", (1960), 17, 184. Moore (1979), 110, mentions Lankester renouncing Anglicanism, and other background details can be found in Goodrich (1930). Lankester's anti-teleology is emphasised in Heron-Allen (1929), 364.

44 Lankester (1880), 26–30, 32–3, 58–62; Lankester (1877), 437–8, 439–440; and see also his *Encyclopaedia Britannica* article in (1890), 349–350. Degeneration was a theme explored by another of Huxley's intimates, Anton Dohrn, one of Haeckel's first pupils and a frequent visitor to the Huxley household. Lankester worked alongside Dohrn at the Biological Station at Naples in 1871–2 and dedicated *Degeneration* to him.

45 D. S. L. Cardwell (1972), 111–126; and especially Haines (1969), Chapter 3 "The Professionalization of the Sciences"; Roderick & Stephens (1972), 7–22. MacLeod (1971) has shown that demands for state salaries and wages for labouring in the national interest grew more frequent during the 1870s and climaxed about 1880 after which a reaction set in. My understanding of the "publicist's" role has been increased by Turner's recent papers (1978, 1980) and he discusses the reaction to scientific naturalism in his book (1974). Lankester's politics would seem to be of relevance regarding his demand for a fair day's wage. According to *The Times*, Mitchell (1929), he showed undeviating allegiance to "Victorian Gladstonian Liberalism". He wanted to see the armed services pruned, a slow extension of the franchise, and the Empire maintained, not by extending its borders, but by "increasing the happiness, comfort, and moral and material welfare of the inhabitants". Interestingly, he retained connections with the Marx

family, whom he visited in a social capacity, and suggested doctors for both Karl Marx and his mother, a fact pointed out to me by Pamela Robinson: Marx & Engels (1943), IX, 388, 389, 414.

46 Huxley (1860b), 393.

47 Think of Tyndall's clarion call at the 1874 Belfast meeting of the British Association: "We claim, and we shall wrest from theology, the entire domain of cosmological theory" – which caused one ungrateful merchant to urge the Home Secretary to prosecute him for blasphemy, Tyndall (1874), 199; Eve & Creasey (1945), 187. Turner (1978), 368, points out that this estrangement of science and Anglicanism was welcomed by a number of the clergy influenced by the Oxford Movement.

48 Lankester (1883), 89. It is not surprising that the word "scientist" came into vogue in the 1870s, despite being deplored by *The Times* (23 August 1878, p. 10) as "a horrible, but handy, Americanism". Of course, it wasn't; the word had been coined by William Whewell at Cambridge long before (1840, I, cxiii), which only goes to show how alien it must have sounded before the new middle-class movement in science.

49 Lankester (1883), 74. The extent of the influence of German anatomy may be gauged by the fact that in the later 1880s (?1889), of the nine class texts used by Lankester at University College, five were German translations (Klein's *Histology*, Claus's *Histology*, Gegenbaur's *Comparative Anatomy*, Wiedersheim's *Comparative Anatomy of Vertebrates*, and Haeckel's *History of Creation*). The four indigenous works were Huxley's *Lessons in Physiology*, Huxley & Martin's *Elementary Biology*, Quain's *Anatomy*, and Darwin's *Origin*. Information from H. J. Harris (1889?), "Notes on a Course of Lectures given at University College London by E. Ray Lankester", 3 Vols, UCL Strongroom MS. Add. 95.

50 Lankester (1880), 59–62, and (1877), 437.

51 Cope began increasingly to acknowledge degeneration in his later writings, such as "On the Evolution of the Vertebrata, Progressive and Retrogressive" (1885). Moore (1979), 151, noticed that he became "somewhat less sanguine about the likelihood of moral progress" at the time when women, blacks, and immigrants were making an increasing mark on American society, suggesting this may have influenced his ideas. Nonetheless, Lankester was undoubtedly a reinforcing factor, and Cope frequently cited his *Degeneration*: Cope (1885), 322, 333, 337.

4. Social Function and Fossil Form

1 Barnes (1977), 2; Barnes (1974); Barnes & Shapin (1979). Notice, though, that a number of historians are more cautious about this denial of a creative role to psychology, e.g. Rudwick (1980), 278–9 (and note 8) and Wood (1980), 19–20.

2 Barnes (1977), 10; Barnes (1974), vii; Rudwick (1980), 271; R. M. Young (1979) provides a convenient way into the literature, as does Mulkay (1979). Neve's review (1980) makes it plain how welcome these new approaches are.

3 Steven Shapin, "History of Science and its Sociological Reconstructions" (1982, in press).

4 For earlier studies, see Colbert (1971) and Delair & Sarjeant (1975). My paper (1979) goes a little deeper into the scientific issues than the present study.

5 Poore (1901); on Grant's "satirical references to Providence" see Godlee (1921) and Schäfer (1901); the most detailed biography of Grant was published in Wakley's *Lancet*: Anon. (1850).

6 Barlow (1958), 49–51; Schäfer (1901).

7 *The Lancet* (1836–7), I, 766. H. E. Gruber (1974), Chapter 2 and 10, looks at the problems of materialism and Darwin's measures to evade persecution (real or imagined).

8 I should like to thank Richard Freeman of University College London for discussions on Grant's suspected homosexuality.

9 *The Lancet* described Grant as the new Cuvier in (1836–7), I, 21. Bourdier (1969) deals with Geoffroy's palaeontological evolution.

10 Desmond (1979), Bowler (1976), 86–7; Owen (1841), 191–204; and on "degradation" in general, Gillispie (1959), Chapter 6.

11 Beddoe (1910), 33. The remaining Grant letters at University College are mostly loan requests.

12 E. Forbes to Huxley, 16 November 1852, IC 16.170; also Huxley (1887a), 188.

13 L. Huxley (1900), I, 303. F. Gregory (1977), and Mendelsohn (1977), 8–11, link the mechanistic trends in German biology with the materialism of the political radicals.

14 Turner (1978), 363; Turner (1980); MacLeod (1971).

15 Huxley (1870b), 191; Huxley (1874a); Tyndall (1867), 92–3, and (1863), 441. Mivart (1871), 18, replied that theologians would not "be in the least painfully affected" by the sight of scientists putting together a fish from inorganic chemicals. For Huxley's "Physical Basis of Life" see Geison (1969) and Huxley (1868d).

16 Cockshut (1964), 91; G. M. Young (1977), 116; Brown (1947), 139; Baynes (1873), 505–6.

17 As the *Edinburgh Review* acidly commented: Baynes (1873), 505–6. For Huxley's belief that materialism contained a "grave philosophical error" see L. Huxley (1900), I, 224, 229, 242–4, and Huxley (1868d), 155. Tyndall (1870), 163–4. On Tyndall's life see Eve & Creasey (1945) and Huxley (1894a).

18 Tyndall (1877), 374; Mivart (1878), 228. For a discussion, see the early chapters of Greta Jones' *Social Darwinism and English Thought* (1980).

19 John Evans (1865), 421; A. Wagner (1862), 266–7 (F. Gregory [1977], deals with Wagner's dispute with Vogt); Mackie (1863); H. Woodward (1862).

20 H. Woodward (1862), 318, 319; Mackie (1863), 2, 4–5; Owen (1863a), 46; Meyer (1862), 366–370. Hooker, having "heard a fraction of Owen's paper on the *Gryphosaurus* at the RS", was also aware of the doubts. "The general opinion was that Owen demonstrated its ornithic affinity and proved it to be a bird with the tail-feathers set on a jointed tail . . . but some say that there are peculiar bones or organs . . . that may yet prove it to be Reptilian. The most curious part of its history is its confirmation of Darwin's much disputed dogma, the 'imperfection of the geological record'." L. Huxley (1918), II, 32.

Like Owen, a number of older geologists accepted *Archaeopteryx* as a primitive bird for their own reasons which owed nothing to Darwinism or evolution. Edward Hitchcock was convinced the "feathered fossil" was a bird because it reinforced his faith in the *avian* nature of his track-making Connecticut Valley creatures: Hitchcock (1863). James Dwight Dana, on the other hand, needed a lowly bird for taxonomic reasons. Each vertebrate class in his classification scheme "consists of, *first*, a grand *typical* division, embracing the majority of its species, and *secondly*, an inferior or *hemitypic* division, intermediate between the typical and the class or classes below": Dana (1863), 316. Typical mammals were complemented by inferior marsupials, typical reptiles by inferior amphibians – and now the more typical of the Aves were counterbalanced by "reptilian Birds". But he was adamant that *Archaeopteryx* was no vindication whatever of "the Darwinian hypothesis" (321). This shows quite strikingly that the new fossil was not only of use to Darwin, but could be accommodated in a number of rival theoretical structures.

21 Joan Evans (1943), 115–6, mentions Lyell and Falconer, and see also Lyell (1863), 453. In her biography, Joan Evans dwells on Huxley's fondness for her father (who "knows the wickedness of the world and does not practise it" – and what better commendation coming from Huxley?). Mackie and Blake can be found in Mackie (1863), 7; and

John Evans (1865) quotes the von Meyer letter. Despite support for Evans, there remained grave doubts over whether these really were *Archaeopteryx*'s teeth.

22 Huxley (1868b), 340.

23 Huxley (1867a), 241; Seeley (1866), 326; Huxley (1868a) and (1864a), 69. W. K. Parker, in his study of the reptilian affinity of *Archaeopteryx*, was already using Huxley's term "Sauropsida" by 1864: (1864a), 57.

24 Huxley (1877), 58–9; Huxley (1868b), 345; H. Woodward (1862), 319. Huxley was not blind to the singularities of *Archaeopteryx*, and in his Royal Institution lecture (1868a) pointed out that "in certain particulars, the oldest known bird does exhibit a closer approximation to the reptilian structure than any modern bird." A number of historians have latched on to this statement, forgetting Huxley's qualifier: that this approximation was only in respect of the tail and hands – "the leg and foot, the pelvis, the shoulder-girdle, and the feathers . . . are completely those of existing ordinary birds." They also tend to ignore the wider canvas – including Huxley's neo-Lyellism – which gives his words a less familiar meaning.

25 Hitchcock (1863), 46, and (1848), 250–1; H. Woodward (1862), 313–5; Huxley (1868c), 366, and (1869a), 388. See Marsh (1877), 11, for reservations. Dean (1969) gives an account of the discovery of the Connecticut Valley tracks, and Guralnick (1972) discusses Hitchcock and his religion.

26 Huxley (1868a). Incidentally, he was later (1877), 65, very cleverly to broach the subject of feathers on *Compsognathus*, though admitting that there was no evidence for this. But even to raise the doubt had a tactical advantage, given his views on the avian looks and physiology of the dinosaur.

27 Huxley (1870a), 532 and (1877), 67.

28 Schmidt (1874), 265; Marsh (1877), 17, 20; Cope (1867); C. Darwin (1872), 302.

29 Phillips (1871), 196, and his letter to Huxley in Huxley (1870d), 466–470. Mantell's appraisal is to be found in Curwen (1940), 96, and on Phillips's ability to keep pace, John Evans (1875), xxxvii. His mature views on descent are to be found in (1871), 400–407. See Phillips (1860b), xxxi, xxxvi–ii, and (1860a), 204, 216 for his immediate reaction to attempts to "tear aside the veil". His views were privately discussed by Darwin in F. Darwin & Seward (1903), I, 130 and II, 31, and in F. Darwin (1887), II, 358. On Phillips' life and work see Anon. (1870a), 301–6 and *Geological Magazine*, 1 (1874), 240. Just as he proscribed evolution, so in 1859 Lyell suggested to Huxley that it might only apply to species, while larger groupings like Classes and Kingdoms, which seem "fixed & absolutely limited", might be the

result of some "as yet undiscovered modus operandi", which we may designate "'creation' meaning thereby that it has not yet been brought within the domain of science", Lyell to Huxley, 17 June 1859, IC 6.20, cf. Wilson (1970), 261.

30 H. Woodward (1893), 52; H. Woodward (1874), 8, 13; Anon. (1921), 483.

31 Hulke (1871), 200 and "Discussion" 206; Hulke (1874b), 26 and (1876), 364–6. Anon. (1895), 510–511, refers to him as a "first authority"; see also H. Woodward (1896), lvi. Hulke otherwise seems to have kept cordial relations with Owen – or, at least, his letters had a friendly air. For example, in 1874 he thanked Owen for his monograph (received via Enniskillen) and offered information on Fox's fossils in return. See J. W. Hulke to R. Owen, 25 June 1874, BM(NH), OC 15, f. 456. This collection also contains many of Fox's letters to Owen (vol. 13, ff. 13–42).

32 Seeley (1872), 42–3.

33 "Discussion" of Huxley (1870d), 486.

34 J. W. Gruber (1960), 104; Mivart (1897), 988–9; Vorzimmer (1972), Chapter 10.

35 St G. Mivart to C. Darwin, 10 January 1872, CUL DAR. 171 (a copy is housed in the BM[NH]).

36 Mivart to Darwin, 25 April 1870, ibid. The situation could only have been aggravated by the English translation of Vogt's inflammatory *Lectures on Man* (1864) and Haeckel's crusading *Generelle Morphologie* (1866).

37 F. Darwin & Seward (1903), I, 336; C. Darwin (1872), 385; Lankester (1870), 34–5. All obiturists agreed with E. S. Goodrich (1930, xiv) that Lankester's "impetuous temperament sometimes led him into difficulties, and even injured his worldly prospects." According to *The Times* (Mitchell, 1929) "He had almost a genius for putting himself in the wrong by explosive and unconsidered action in a just cause". Lankester fought law suits, university governing bodies, even the Piccadilly police. As a consequence he was never elected President of the Royal Society (when a perfect choice) and was compulsorily retired as Director of the Natural History Museum. But in some ways this only made his presence in late Victorian and Edwardian biology the more commanding. His scope and powers were prodigious, and his researches – whether on Devonian jawless fish, embryology, or parasites – monumental and enduring, although in the current historiographical climate it is as a propagandist that Lankester becomes most interesting (Chapter 3).

38 Mivart (1870), 113–5, 117, 118; and (1893) for his appreciation of Owen.

39 Flower (1870); Mivart (1869).
40 I have not seen Darwin's letter, but Mivart's reply begins "I herewith
close this correspondence & will say nothing even in this letter
calculated to annoy you in the least. Perhaps it would be better taste not
to reply *at all* but as you let me know you will not read my reply to Mr
C. Wright . . ." Mivart to Darwin, 10 January 1872, CUL DAR.171.
On Huxley's reaction see Mivart (1897), 994–9; J. W. Gruber (1960),
84 et seq; Bartholomew (1975), 532–4; Vorzimmer (1972), 244–250;
Moore (1979), 62–4, 117–122.
41 Mivart (1871), 67, 70–3; on the "grove of trees" see Mivart (1873), 510
and J. W. Gruber (1960), 33; and on "chance" J. B. Mozley (the
Quarterly Reviewer) (1869), 170.
42 J. W. Gruber (1960), 87. For Huxley's dislike of Catholicism, see
Moore (1979), 63–4, and Huxley (1871a), 147; and on Hooker's
"blackballing", Ruse (1979), 255.
43 St G. Mivart to R. Owen, 26 October 1871, BM(NH), OC 19.267.
44 Bibby (1960), 184. Ray Lankester actually took the Chair of Zoology
at University College in February 1875 (not in 1874 as many obiturists
claim) – William Henry Allchin (1846–1911) having filled in after the
death of the 80-year-old Grant in August 1874.
45 R. Owen to J. Tyndall, 14 June 1871, BM(NH) OC 21.28; Eve &
Creasey (1945), 159, quote this reply incorrectly.
46 Huxley (1870c) and (1870e), 487–9; Hulke (1873), 530, and Owen's
"Discussion", 531–2; Hulke (1874a), 22–3. Hulke found the dinosaurs
"peculiarly interesting" (1882, 1035) partly "on account of their
forming a link between more specialised Reptiles and Birds."
47 Owen (1874–1889), 87–93, which is the main source for Owen's
reaction, but see also (1872–1889), Supp. 4, 14–15, and Supp. 5, 15–18,
and (1878), 426.

5. "Phylogeny"

1 Challinor (1971), 139.
2 The handover of power after the Civil War from Joseph Leidy to the
committed evolutionists O. C. Marsh and E. D. Cope would make a
valuable study. (Leidy complained that he was forced out of the field
by their "long purses" but the issue runs far deeper than this – Geikie,
1892, 58.) Gerstner (1970) has looked at early Transatlantic palaeontology (c. 1830s); J. T. Gregory (1979) provides a bibliographic sweep
from 1776–1976; and Rainger's essay (1981) covers Marsh, Cope, and
their students in the period 1880–1910. Any study of the transition

ought really to relate the emergence of the 'American school' of palaeontology and evolution to post-war social and industrial conditions. A number of scholars (Pfeifer, 1974; Aldrich, 1974) have covered the Darwinian controversies both within and without the scientific community. More generally, Social Darwinism has been treated by Hofstadter (1955) and Russett (1976).

3 Phillips (1871), 400–405.

4 Allen (1978b), 205; Porter (1978), esp. 829–830. Allen's *Naturalist in Britain* (1978a) looks in detail at the many amateur and professional Victorian societies. Cope talked of Darwin leaving Hamlet out of the play in (1880), 225.

5 L. Huxley (1900), I, 267. Haeckel's "phylogeny" was structurally distinct from pre-Darwinian developmental theories: Hodge (1972) points out that the *Vestiges* lacked any notion of *common* descent and rested on a multitude of parallel lines, each stretching towards the human apex – see also Winsor (1976), 19, and Ruse (1979), 10, for diagrammatic representations of Chambers' views.

    The recent biogeographical studies in the *Journal of the History of Biology* are Richardson (1981), Nelson (1978), Fichman (1977). Brockway (1979) has opened up the subject of the scientists' role in colonial expansion, although her book leaves scope for more detailed historical work in this important area.

6 Phillips (1860b), xxxi; Burkhardt (1974), 72–3, discusses the learned societies. For a review of *Sacred Steps*, see Anon. (1865). Sedgwick's inductivist criticisms are to be found in F. Darwin (1887), II, 248, and Sedgwick (1860), 285.

7 Huxley (1869c), 416; Huxley had begun to explode the Baconian myth as early as 1860 in his *Westminster* review of the *Origin* (1860a, 72–3), although nowhere is he more vehement than in his working men's lectures, published in 1862 under the title *On our Knowledge of the Causes of the Phenomena of Organic Nature* (1862b), 56. For Tyndall's "Scientific Use of the Imagination" see (1870), 130, 158. Huxley actually talked of "my scientific young England" in a letter to Hooker in 1858, discussing "that madcap Tyndall" (like Huxley an Alpine climber) breaking his "blessed neck some day" and leaving "a great hole in the efficiency of my scientific young England": L. Huxley (1900), I, 160. Tyndall, Hooker and Huxley were shortly to band themselves together as the *x* Club.

8 Bölsche (1909), 5.

9 Haeckel (1876), I, 1–2; for details of his sales, see the introduction to Haeckel (1904), vi. Both F. Gregory (1977), 183, and Bölsche (1909), 6, mention Büchner's title for Haeckel, the "German Darwin". Nordenskiöld made his claim in (1942), 515. Rinard (1981) discusses the

development of Haeckel's 'Biogenetic Law', and Kelly (1981) the popularisation of Darwinism in Germany.

10 Gasman (1971), 18. For my interpretation of Haeckel I have leaned heavily on Gasman's important book. For Marx's comment on Darwin's "crude" style, see Marx to Lassalle, 16 January 1861 in Marx & Engels (1943), IX, 125–6.

11 Haeckel (1876), I, 11.

12 Haeckel (1876), I, 4; the "noblemen and dogs" jibe is on p. 295.

13 Gasman (1971), 47, 111.

14 Gasman (1971), 18, 17.

15 Haeckel (1876), I, 172, 277–9; Haeckel (1866), II, 257; Gasman (1971), 126–8. Haeckel upheld the Spartan ideal of smothering weakly infants and deplored the military practice of sending the fittest young men to the front, while leaving the retarded at home to "propagate".

16 L. Huxley (1900), I, 288; for Darwin's "groaning and swearing", F. Darwin & Seward (1903), I, 274; Bölsche (1909), 94.

17 See McCabe's note, page 75, in his translation of Bölsche (1909). Haeckel's "pious inquisition" can be found in (1876), I, 19, and Vogt's "Apostle skulls" in Vogt (1864), 378. See F. Gregory (1977), Chapter 3, for a study of Vogt.

18 F. Darwin (1887), III, 104; F. Darwin & Seward (1903), I, 277–8; L. Huxley (1900), I, 305. Haeckel was still hoping for an English translation as late as 28 February 1869, when he sent Vol. I of *Generelle Morphologie* with all the necessary corrections and modifications to Huxley (IC 17.198).

19 F. Darwin & Seward (1903), II, 350; L. Huxley (1900), I, 266; Haeckel (1876), II, 248.

20 L. Huxley (1900), I, 299; for Huxley's reservations on philosophic materialism, see (1868d), 154–165; and his review of Haeckel, Huxley (1869b), 13–14, 43. For the private reactions of an intransigent idealist like Louis Agassiz to Haeckel's *History of Creation*, and what he considered the misuse of many of his own discoveries, see Gould (1979).

21 F. Darwin (1887), III, 105; Haeckel (1876), II, 38. Gould (1977), 170–3, discusses Haeckel's manufacture of ancestors from embryological models.

22 Huxley (1869b), 41; Schmidt (1874), 250.

23 Haeckel (1876), I, xiii; cf. Huxley's letter in *The Ibis*, 4 (1868), 357–61, where he states his aim is a *"genetic classification"* (361). The family tree of the Gallo-columbine birds is contained in his (1868c), 365. For his apologies on not reading Haeckel's book, L. Huxley (1900), I, 288. Huxley, by the way, was in constant touch with Haeckel over his new avian "genetic" classification and attempts at phylogeny, see, e.g. Haeckel to Huxley, 27 January 1868, IC 17.183.

24 L. Huxley (1900), I, 303.

25 Helfand (1977), 161.

26 Huxley (1865a), 29; Huxley (1866), 14. Paradis (1978), 31, 36, discusses the "proletarian roots" of Huxley's scientist. Huxley called himself "a plebeian who stands by his order" in (1874b), 189.

27 Paradis (1978), 66–7; for another approach to the changes in Huxley's world-view, see Eng (1978). L. Huxley (1900), I. 282 deals with the Eyre affair.

28 Huxley (1868e), 42.

29 Huxley (1887d), 108; cf. Paradis (1978), 66–7. G. Jones (1980) discusses Wallace and social theory (pp. 24–34) and socialism in Chapter 4, as well as the use of Darwinism to liberal radicals in further reducing aristocratic privilege and state controls on economic life (Chapter 3).

30 Huxley (1871b), 256; Helfand (1977), 160.

31 Huxley (1863a), 111; figures for his annual earnings taken from Bibby (1960), 53 and (1972), 81. The comment by *The World* reporter is cited in L. Huxley (1900), I, 463.

32 R. Godwin-Austen to Huxley, 30 March 1863, IC 10.183; D. S. L. Cardwell (1972), 71–3, discusses the lower middle-class invasion of the Mechanics' Institutes, although Becker (1874), 186, sat in on one of Huxley's Working Men's Lectures in the early 1870s and reported that it was crammed with men fresh from the workshop – and that for once their place had not been "usurped by smug clerks or dandy shopmen".

33 Huxley (1870a), 529.

34 "Just as there were no political revolutions in England in 1789, 1830, 1848, and 1871", says Stebbins (1974), 163. For a study of the institutional differences between the science of prehistory in England and France, see Boylan (1979).

35 Lyell (1868), II, 481–4; Darwin's letter to Gaudry is in F. Darwin & Seward (1903), 235–6; and Huxley praised Gaudry's opus (Gaudry 1862–7) in Huxley (1870a), 530.

36 Rütimeyer (1868), for his work on *Bos* and the derivation of cattle from Pliocene forms, see Rütimeyer (1867–8), and for an overview, Rütimeyer (1867).

37 Huxley (1870a), 532–3; he talks of "retrospective prophecy" in (1880c).

38 Owen (1851), 449–450; Ospovat (1981), 137–140 makes exactly the same point about Huxley's inability in post-Darwinian times to get much beyond Owen. For Huxley's denial that *Palaeotherium* was "a more generalized type" any more than a father is more "generalized" than his sons, see Huxley (1860b), 389 – a statement which he had all but retracted in (1870a), 535, when he talked of derivation from the

"average" form. Owen championed the "derivation of equines" in (1866–8), 791–6, and cf. his fig. 614 (p. 825) with Huxley's more inclusive illustration in *American Addresses* (1877), fig. 9, p. 88. Marsh's Yale expeditions to the Western Territories (1870–3) were so successful that in 1874 he declared that the horse "line in America was probably a more direct one, and the record is more complete [than in Europe]": Marsh (1874), 255. Three years later he was able to pronounce America "the true home of the Horse": (1877), 27. Simpson (1951), 87, suggests that Huxley was first to put together the horse lineage, actually he arrived on the scene rather late, Owen, Gaudry, *et al* having done much of the ground work.

39 Rudwick (1972), 262; Foster (1869), 397.

40 Vucinich (1974), 228; Todes' essay on Kovalevskii's revolutionary activity and palaeontology (1978) is the most thorough to date. Kovalevskii (1873), 21, acknowledged the "complete revolution" caused by the *Origin*. Politics in Tsarist Russia were synonymous with radical activity for, as Rogers (1973), 502, says, no conservative school "could even conceive of using Darwinism to rationalize a social structure resting upon an autocracy which found its ultimate justification in divine sanction". It should, perhaps, be pointed out that a good part of Kovalevskii's work was complete by the time he reached Jena in December 1871. As for Haeckel, Stauffer (1957), 139, states that despite coining the word, he "did not display any notable insight into the dynamic principle of ecology".

41 Flower (1873), 101.

42 Flower (1873), 104. On transposed concepts in Lyell's geology, and particularly his analogy with historiography, see Rudwick (1978b).

43 This statement was repeated on three occasions to my knowledge: in Huxley's original lecture on "persistent types" (1859c), 92, in his catalogue of fossils in the Jermyn Street Museum (1865b), 175 – apparently written before he had read the *Origin* – and in the 1862 address (1862a), 524. Compare his views in the later address (1870a), 538.

44 From *The Moray and Nairn Express* obituary of Gordon: Anon. (1893). Copies of a number of obituaries from local papers were kindly sent me by Mike Benton. Huxley's study of *Stagonolepis* (1859d) apparently started him on his crocodile investigations. He about-turned and argued for progression in (1875). On the material being dispatched to him in the meanwhile, see G. Gordon to T. H. Huxley, 10 August 1866, IC 17.83.

45 Huxley (1876), 175. Darwin in the *Origin* had grave doubts about finding any rational criterion for "highness", doubts echoed by

Michael Foster in his article "Higher and Lower Animals" (1869). Huxley was astonished at the 'persistence' of *Telerpeton* in (1876b), 213, although writing in *Nature* on the origin of vertebrates (1876), 173–4, he did concede that older lizards had "a slightly simpler structure" than living ones.

46 See the comments by Duncan, John Evans and Harry Seeley in the "Discussion" following Huxley's paper (1875). It was here that Evans inquired about the migration of the internal nostrils being related to a new mode of life. Osborn in *The Age of Mammals* (1910), 6, 8, talks of the "new spirit" in Kovalevskii's papers.

47 Owen (1871), 108, 110; Owen (1860a), 314, for the crack about "congenital" blindness. Hulke introduced the Metamesosuchia in (1878), 380–1.

48 Marsh (1879), 21–2; Haeckel (1876), I, 54, 59, 126; L. Huxley (1900), II, 39. Huxley had specifically singled out Cuvierian "rational correlation" for criticism as early as (1856b). Even Kovalevskii made some unkind cuts: Todes (1978), 140–1.

49 Owen (1878), 421. The dwarf crocodiles were introduced in Owen (1879), 149. For the criticism of his failure to mention "evolution", see the "Discussion" following his 1878 paper, 428–430.

6. Groves of Trees & Grades of Life

1 Huxley's campaign against the reification of laws and causes went unabated from the fifties when he attacked Chambers and Owen (Huxley, 1854, 1858) until the time he took Canon Liddon and the Duke of Argyll to task for indulging in "Pseudo-Scientific Realism" (Huxley, 1887b, 1887c). Moore (1979) has produced an extensive critique of the "military metaphor" and the milieu in which Draper and White wrote their 'warfare' histories. The latter has recently been abridged and reissued (White, 1965).

2 Mivart (1871), 238; John Herschel reputedly called Darwin's selection "higgledy-piggledy", see F. Darwin (1887), II, 241.

3 Peckham (1959), 188–190. Argyll (1906) describes his life in politics, and Gillespie (1977) his debate with Lubbock at this time. R. Smith (1973), esp. 97–105, has used contemporary psychological theories to elucidate the growing distinction between the positivists' use of a Humean "constant conjunction" theory of causation, and the older belief in an "efficient" cause equivalent to a "force", i.e. which required a certain Divine effort. Mivart used his crystal analogy in (1871), 143.

4 R. Smith (1974), 279–80. On the "Cambridge Network" see W. F. Cannon (1964), reworked in S. F. Cannon (1978), Chapter 2; and on the "scientific naturalists", Turner (1974). Huxley mentions his and Tyndall's political differences in L. Huxley (1900), I, 283. Smith was reviewing Hull's *Darwin and His Critics*, although very few historians have taken a "community" approach. The quote on Argyll's "eroded" Calvinism comes from Moore (1979), 343.

5 Argyll (1868), 259–60; Gillespie (1979), 93–104; Argyll (1906), II, 498, employs the embryological analogy, J. B. Mozley (1869), 162, whom Chadwick (1975), 41, calls the "theologian's theologian".

6 T. H. Huxley to J. Knowles, 9 March 1887, Wellcome Institute AL 67980; see Huxley (1887b, 1887c). Darwin's comments on the Duke are to be found in F. Darwin (1887), III, 61. Being a positivist, of sorts, Huxley held a constant-conjunction theory of causation. This was the position from which he attacked the Duke.

7 R. Owen "The Reign of Law" (1867), *Autograph Manuscripts of Sir R. Owen*, BM(NH), OC 59, pp. 1, 2, 7, 18, 24.

8 Oppenheimer (1967), 318; Bowler (1977), 37, suggests that Owen had "abandoned his original position by this time, recognizing on the basis of his palaeontological studies that divergence and specialization not unity are the key factors in the history of life", but I doubt if the issues were so clear cut in his mind.

9 Thomson (1868), 64; Jenkin (1867), 301; Tait (1869), 438. For commentaries on the age estimates and the relationship of Thomson and his "worshippers", Tait and Jenkin, see Sharlin (1979), Chapters 9 and 10, and Burchfield (1975), Chapters 2–4.

10 St G. Mivart to R. Owen, 8 June 1879, BM(NH) OC 19.261. Butler talked of "sleight-of-hand" in *Evolution, Old and New* (1879), 346, and of a bishop in preference to Huxley in H. F. Jones (1920), I, 385. For Lyell to Argyll on "creatives power" see K. Lyell (1881), II, 431–2.

11 Jenkin (1867), 294; F. Darwin (1887), III, 107; F. Darwin & Seward (1903), II, 397; Huxley (1871a), 121; Vorzimmer (1972) and for a review of his book, Gould (1971).

12 Merchant (1916), II, 31; R. Smith (1972) gives perhaps the best overview of Wallace's "spiritual interpretation of nature" (186). The sources of the three unpublished quotes are: on the "painful effort", St G. Mivart to C. Darwin, 23 April 1871, CUL DAR.171 (copy in BM[NH]); on the "*odium theologicum*", Mivart to Darwin, 26 January 1871, ibid.; and on the "*internal* principle", Mivart to R. Owen, 26 October 1871, BM(NH) OC 19.267.

13 Mivart (1871), 239, and (1872), 123. Most recently, Ridley (1982) has looked at the idealists' use of coadaptation (or harmonious adjustment) as an argument against natural selection.

14 Dawkins (1865), 414; on Dawkins's life, see Smith Woodward (1930 –1).

15 Dawkins (1868), 438–9, 444–5, quoting from Spencer's *Principles of Biology* (1864), I, 271. Dawkins talks of the "trinity of causes" in (1868), 442. Dawkins did little towards establishing any particular genealogy, and it was not until 1878, e.g., that he mentioned the gradual complication of antlers during Miocene time: Dawkins (1878), 419.

16 Dawkins (1871), 195. *The Times* review of Darwin's *Descent* is 8 April 1871, 5; and details of Owen's Parisian joint-stock holdings can be had from Owen (1872). Mivart's letter to Darwin on "religious decay" is 24 January 1871, CUL DAR.171 (copy in BM[NH]).

17 Dawkins (1871), 201; subsequent quotes from pp. 200, 206–7.

18 Mivart (1873), 506, 508–9; Montgomery (1974), 100–1. The image of the grove of trees is evoked in Mivart (1873), 510.

19 Mivart (1873), 506; J. Gruber (1960), 33. Haeckel comes down on the side of monophyly in (1876), II, 46. Howes (1905), 98, calls Mivart's diphyletic origin of the mammalia "sensational", see Mivart (1888), 376 for his original statement. He was responding to Poulton (1888) and W. H. Caldwell (1887), and Poulton's rejoinder came in (1888–9).

20 For this, and the information concerning Huxley's testimonial, I am indebted to Patricia Methven, Archivist in the King's College Library. Evidently the obiturist in the *Geological Magazine* was incorrect in stating that Duncan took the Chair in 1870: J. W. Gregory (1891), 333 (incidentally it was Gregory who singled out *British Fossil Corals* for such high praise, ibid. 334). Ruse (1979), 259, calls Duncan a "fellow traveler", and indeed in the "Discussion" following Huxley's paper on crocodile evolution in 1875 Duncan was quite complimentary: Foster & Lankester (1902), IV, 82.

21 P. M. Duncan (1865), 361 – although only a few lines further on he talks of an "inherent power of variation".

22 P. M. Duncan (1877), 87; previous quotes from pp. 84–6.

23 Anon. (1891), 388.

24 Seeley (1901), v. My study was facilitated by the use of his two volumes of offprints, bound under the title *Papers by H. G. Seeley*, included among Sir Arthur Smith Woodward's books in the Natural Sciences Library at University College London. From these one can build up the sort of synoptic picture impossible to obtain from isolated papers (also some of the items included are obscure or unlisted in his usual bibliographies). Smith Woodward married Seeley's daughter Maud – on which see his note pasted inside volume two. The papers carry occasional annotations by Seeley, Smith Woodward, and Owen, whose complimentary copy of Seeley's "History of the Skull" – a

pamphlet printed by the Science Society of King's College in 1882 – has found its way back into the collection.

25 Sollas (1909), lxxii. Cope's views on Seeley, Huxley, and Owen are to be found in Osborn (1931), 243, 244, 247, and esp. 253. Although Seeley fails to figure in the *Lives and Letters* of the Darwinians, he was warmly congratulated by Huxley and Tyndall for his Nottingham speech in 1866 on science education in schools (not surprisingly, given Huxley's particular interest in the subject) – see Seeley to A. Sedgwick, 28 September 1866, CUL 7652.II.GG6.

26 The details of Seeley's early life are taken from three letters to Adam Sedgwick: 7 September 1866, CUL 7652.II.GG2; 17 September 1866, CUL 7652.II.GG3; and 24 September 1866, CUL 7652.II.GG5. He mentions his father's bankruptcy and being one of the "great unwashed" in GG2; and his excitable nature and the decline of the family fortunes in GG3 (his grandfather was evidently a Clerk in the Treasury earning £1000 p.a.).

27 Seeley (1861), 4; cf. Forbes on "generic ideas" (1854), lxxx. Seeley mentions preferring Forbes' lectures to Huxley's in GG5 (see note above) and in the same letter discusses the Working Men's College. A. S. Woodward (1910), xv–xvii, also mentions his attendance at Forbes' lectures. Anon. (1907), 242, discusses the benefits of having studied under Owen when he came to deal with the Cambridge fossils. It is perhaps significant that he followed in Owen's footsteps, specialising in pterosaurs and mammal-like reptiles, and although he refused to kowtow to Owen on specifics, there were areas of substantial agreement: e.g. Seeley considered that there might have been 'a common parentage for Dinosaurs and Mammals" (1871), 318, thus fulfilling Owen's ambition of linking the two. Nor did Seeley have much love for "Prof. Huxley's morphological hypothesis of the Avian affinities of Dinosaurs" (1875a), 442. And however much he endorsed Huxley on hot-blooded pterosaurs, he refused to say as much for dinosaurs, siding with Owen and considering them "animals of sluggish habits, and therefore, it may be, of cold blood" (1875b), 149.

28 Wynne (1979), 173–5; Rothblatt (1968). Seeley admitted his ignorance of Spencer's views in (1867), 371, and mentions taking his stand "not as a working man" in GG2 (note 26 above).

29 A. Sedgwick to R. Owen, 30 March (no year), BM(NH) OC 23.298. For Seeley's views on pterosaurs in the mid-sixties see (1864), 69, and (1866a), 326.

30 Seeley to Owen, 28 July 1869, BM(NH) Owen Collection 14 (inside Owen's copy of *British Fossil Reptiles*).

31 Owen (1866–8), I, 6, 18; cf. his derivative approach (1874–89), 91. When Cope (1871) erected his "Archosauria" in 1868–9, he disting-

uished it from the "Pterosauria" on osteological rather than physiological grounds. To what extent he was influenced by Seeley I do not
know.

32 Owen (1869), 72–74, 80. On pp. 74–5 Owen states his reasons for
favourably comparing birds and mammals, which share characters
concerned with the "higher phenomena of life" – intelligence, cardiovascular physiology, blood temperature, etc. – using this "grade"
concept to discredit Huxley's sauropsid bird-reptile connection: i.e. he
is replying to Huxley (1867), 241, and (1868a), as well as Seeley
(1866a). Seeley's retort, in which he calls Owen "unjust", is his
(1870a). When Robert Lee, a Cambridge contemporary, received an
offprint of this paper, he told Seeley that he was gratified to see Owen
"handled fairly for he has played foul all through life, led by ambition
rather than love of truth & [is] now receiving the verdict of posterity
rather sooner than expected": R. J. Lee to H. G. Seeley (n.d.), BL
Add. 42,585, f. 38. I would like to thank Kevin Padian for his
constructive criticism of my analysis of the Owen–Seeley relationship,
and for showing me his MS, "Historical Misconceptions of Pterosaurs
and the Owen–Seeley Dispute", which constitutes Chapter 3 of his
Ph.D. "Studies of the Structure, Evolution, and Flight of the Pterosaurs", Yale University, 1980.

33 Huxley (1868a), 308; cf. the *Annals* review of Seeley's book (Anon.
1870b, 188) which discusses Huxley and Owen on "hot blood" and the
latter's recourse to non-physiological "adaptive" characters. Seeley
(1876), 85–7 answered his critics and insisted that the pterosaur lungs,
brain and heart were not "adaptive" at all but part of an avian
birthright.

34 Seeley (1870b), 94; Anon. (1870b). Both in *The Ornithosauria* and in the
"Discussion" following Huxley's (1870d) did Seeley attempt to dissociate dinosaurs from reptiles: *Quart. J. Geol. Soc.*, 26 (1870), 31, for
his belief that they were as "unlike reptiles as birds or mammals".

35 Seeley (1872), 42–3, 44; he talks of Owen's "beautiful" description,
and the "brain-case" being "a modified vertebra" in (1866b), 345, 362;
and of morphology as the future "lawgiver" in (1882), 14. Cf. Cope
(1887), 181, who also considers the "vertebral skull" a complex and
unresolved issue.

36 Seeley (1887).

37 Huxley (1879), 354, 355; the "title-deeds" were mentioned in (1868a),
304, and Rudwick (1972), 250, comments on this appropriate analogy
for property-conscious Victorians.

38 Bowler (1976), 134; and on Seeley deciding to extend Owen's work,
A. S. Woodward (1910), xvii. For his personal assessment of the South

African "mission" see Seeley (1889), and for some letters, Swinton (1962).

39 A. G. Bain to Rev. J. Adamson, 22 April 1851, BM(NH) OC 2.35. *Dicynodon* was set off in the order Anomodontia by Owen, and in 1876 on erecting the order Theriodontia, he commented on "the mutual affinity of the two extinct groups, and their common possession of a mammalian structure": Owen (1876c), 362. Thus when he eventually speculates on the origin of the Platypus ([1880], 423), he actually derives it from anomodonts.

40 A. G. Bain to R. Owen, 25 September 1848, BM(NH) OC 2.32. Owen mentions the difficulty of separating bone from rock in (1876a), iii; for other details see Bain's own account: (1854), 53–9.

41 Quoted in Swinton (1962), 36; on Prince Alfred's dicynodont, see Owen (1862). Owen comments on *Galesaurus* and its "approach to the mammalian class" in (1860c), 60.

42 Owen included in his definition of the Theriodontia what he considered to be mammalian osteological features, e.g. the loss of the ectopterygoid bone and appearance of the marsupial entepicondylar foramen, Owen (1876a), 75, (1876b), 98–9. Cf. Seeley's discussion of Owen's various papers, *Quart. J. Geol. Soc.*, 32 (1876), 362; 35 (1879), 199; 36 (1880), 425; 37 (1881), 270.

43 T. Bain to R. Owen, 15 February 1878, BM(NH) OC 2.61. On the subject of Bain's remuneration, see Owen (1876a), iii–iv, and *Quart. J. Geol. Soc.*, 35 (1879), 199. Anon. (1907), 244, mentions most of the mammal-like reptiles having gone to Owen before Seeley's arrival.

44 H. Woodward (1893), 52. That Huxley did on occasion examine Bain's fossils in the British Museum, see Huxley (1868f), 298–9.

45 Owen (1876a), 76; and also *Quart. J. Geol. Soc.*, 32 (1876), 99–101. For the Presidential assessment of Owen's papers, see John Evans (1876), 112.

46 Owen (1881), 265.

47 Brown (1947), 139.

48 Huxley (1880b), 468 – Osborn (1910), 31, considered this essay of Huxley's "brilliant".

49 Huxley (1877), 67. For Baur's ability to accommodate what he called the "sauromammalia", see (1887), 102. Here Baur admitted, in his characteristic broken English, that theriodonts were "already too much specialized to be the ancestors of Mammals", so he made them little-modified descendants of the hypothetical Mid-Permian "sauromammalia", which themselves *did* give rise to mammals. Little modified was the crux. Shortly before he died insane, Baur told Seeley that the South African "mammal-like reptiles, or reptile-like mammals, are very important finds. They are certainly not very far from the 'Sauro-

mammalia' as I have called the hypothetical ancestors of the mammals." G. Baur to H. G. Seeley, 17 May 1895, BL Add. 42, 585, f. 66.

50 Marsh (1880), 238–9; cf. (1878), when he still thought *Dryolestes* an opossum-like marsupial. Seeley (1879), 462, talks of a "generalized order". So although Huxley in 1862 had challenged Owen to say in what way a fossil *Phascolotherium* was more generalised than a living opossum, by the 1870s few doubted that this was the case. On the *Dimetrodon*-amphibian connection see Cope (1882), 38–40; on *Dimetrodon*'s discovery, Cope (1878), esp. 529; and on Cope's recognition of Huxley's difficulties, (1884), 480.

51 Seeley (1891), 238. E. S. Goodrich (1916) completed the symmetry of the system by labelling the branch from amphibians to mammals the "Theropsida" (to complement Huxley's reptile-bird branch, the "Sauropsida").

52 For a sensitive and detailed critique of Marxist historiography, as applied to the history of science, R. M. Young (1973). Turner (1978), 369, lists Owen as a "preprofessional".

53 Owen actually classified one theriodont, *Tapinocephalus*, as a dinosaur, (1876a), 1–15, and (1876d), 43–6. This was not deliberate obfuscation, but it does suggest that he was conditioned to find mammal-like features in reptiles. In fact, in the "Summary" of his *Catalogue* (1876a), Owen juxtaposes dinosaurs and theriodonts in an effort to prove the reptilian likeness to mammals.

54 Owen (1880), 423. Here Owen looked on anomodonts as mammalian forerunners, although Cope (1884) derived mammals from carnivorous pelycosaurs (i.e. *Dimetrodon*'s group).

Bibliography

Agassiz, L. (1843), "A Period in the History of our Planet", *Edin. New Phil. J.*, 35: 1–29.

—— & E. Agassiz (1868), *A Journey in Brazil* (New York, Praeger, 1969 reprint).

Aldrich, M. L. (1974), "United States: Bibliographical Essay", in Glick (1974), 207–226.

Allen, D. E. (1978a), *The Naturalist in Britain: A Social History* (Harmondsworth, Pelican).

—— (1978b), "The Lost Limb: Geology and Natural History", in Jordanova & Porter (1978), 200–212.

Anon. (1850), "Robert Edmond Grant", *The Lancet*, II: 686–695.

—— (1860a), "Science: Palaeontology", *The Athenaeum*, No. 1693: 478–9.

—— (1860b), "Professor Owen on the Origin of Species", *Saturday Review*, 9:573–574.

—— (1863), "A Sad Case", *Public Opinion*, 2 May: 497–498.

—— (1864), "Instances of the Power of God", *Geol. Mag.*, 1: 82–3.

—— (1865), "'Scripture' Geology", *Geol. Mag.*, 2: 417–420.

—— (1870a), "John Phillips", *Geol. Mag.*, 7: 301–306.

—— (1870b), "The Ornithosauria", *Ann. & Mag. Nat. Hist.*, 6: 186–188.

—— (1891), "The Late Prof. Martin Duncan", *Nature*, 44: 334–335.

—— (1893), "Death of the Rev. Dr Gordon", *Moray & Nairn Express*, 16 December.

—— (1895), "John Whittaker Hulke", *The Lancet*: 510–511.

—— (1907), "Eminent Living Geologists: Professor H. G. Seeley", *Geol. Mag.*, 4: 241–253.

—— (1921), "Henry Woodward", *Geol. Mag.*, 43: 481–484.

Argyll, Duke of (1868), *The Reign of Law* (London, Strahan), 5th ed.

—— (1869), *Primeval Man* (London, Strahan).

—— (1906), *Autobiography and Memoirs* (London, Murray) 2 Vols.

Ashton, R. (1980), *The German Idea: Four English Writers and the Reception of German Thought 1800–1860* (Cambridge University Press).

Babbage, C. (1838), *The Ninth Bridgewater Treatise: A Fragment* (London, Cass, 1967 reprint).

Bain, A. G. (1854), "On the Discovery of the Fossil Remains of Bidental and other Reptiles in South Africa", *Trans. Geol. Soc.* (London), 7: 53–59.

Barlow, N. (1958), *The Autobiography of Charles Darwin, 1809–1882* (New York, Norton).

Barnes, B. (1974), *Scientific Knowledge and Sociological Theory* (London, Routledge & Kegan Paul).

—— (1977), *Interests and the Growth of Knowledge* (London, Routledge & Kegan Paul).

—— & S. Shapin (1979), *Natural Order: Historical Studies of Scientific Culture* (Beverly Hills/London, Sage).

Bartholomew, M. (1973), "Lyell and Evolution: An Account of Lyell's Response to the Prospect of an Evolutionary Ancestry for Man", *Brit. J. Hist. Sci.*, 6: 261–303.

—— (1975), "Huxley's Defence of Darwin", *Annals of Science*, 32: 525–535.

—— (1976), "The Non-Progress of Non-Progression: Two Responses to Lyell's Doctrine", *Brit. J. Hist. Sci.*, 9: 166–174.

—— (1979), "The Singularity of Lyell", *Hist. Sci.*, 17: 276–293.

Baur, G. (1887), "On the Phylogenetic Arrangement of the Sauropsida", *J. Morphology*, 1: 93–104.

Baynes, T. S. (1873), "Darwin on Expression", *Edinburgh Review*, 137: 492–508.

Becker, B. H. (1874), *Scientific London* (London, Henry S. King).

Beddoe, J. (1910), *Memories of Eighty Years* (Bristol, Arrowsmith).

Bell, S. (1981), "George Henry Lewes: A Man of His Time", *J. Hist. Biol.*, 14: 277–298.

Benton, M. (1982), "Progressionism in the 1850s: Lyell, Owen, Mantell and the Elgin fossil reptile *Leptopleuron* (*Telerpeton*)", *Archives of Natural History*: in press.

Best, G. (1979), *Mid-Victorian Britain 1851–1870* (London, Fontana).

Bibby, C. (1960), *T. H. Huxley, Scientist, Humanist and Educator* (New York, Horizon Press).

—— (1972), *Scientist Extraordinary: The Life and Scientific Work of Thomas Henry Huxley 1825–1895* (Oxford, Pergamon Press).

Blake, C. C. (1863), "Man and Beast", *Anthropological Review*, 1: 153–162.

Bölsche, W. (1909), *Haeckel: His Life and Work* (London, Watts; trans. J. McCabe).

Bourdier, Franck (1969), "Geoffroy Saint-Hilaire versus Cuvier: The Campaign for Paleontological Evolution (1825–1838)", in Schneer (1869), 36–61.

Bowler, P. J. (1974), "Evolutionism in the Enlightenment", *Hist. Sci.*, 12: 159–183.

—— (1976), *Fossils and Progress: Paleontology and the Idea of Progressive Evolution in the Nineteenth Century* (New York, Science History Publications).

—— (1977a), "Darwinism and the Argument from Design: Suggestions for a Reevaluation", *J. Hist. Biol.*, 10: 29–43.

—— (1977b), "Edward Drinker Cope and the Changing Structure of Evolutionary Theory", *Isis*, 68: 249–265.

Boylan, P. J. (1979), "The Controversy of the Moulin-Quignon Jaw: The Role of Hugh Falconer", in Jardanova & Porter (1979), 171–199.

Brockway, L. H. (1979), *Science and Colonial Expansion: The Role of the British Royal Botanic Gardens* (New York, Academic Press).

Brooke, J. H. (1977a), "Natural Theology and the Plurality of Worlds: Observations on the Brewster-Whewell Debate", *Annals of Science*, 34: 221–286.

—— (1977b), "Richard Owen, William Whewell, and the *Vestiges*", *Brit. J. Hist. Sci.*, 10: 132–145.

—— (1979), "The Natural Theology of the Geologists: Some Theological Strata", in Jordanova & Porter (1979), 39–64.

Brown, A. W. (1947), *The Metaphysical Society* (New York, Columbia University Press).

Buckland, W. (1837), *Geology and Mineralogy Considered with Reference to Natural Theology* (London, Pickering), 2 Vols.

Burchfield, J. D. (1974), *Lord Kelvin and the Age of the Earth* (London, Macmillan).

Burkhardt, F. (1974), "England and Scotland: The Learned Societies", in Glick (1974), 32–74.

Burn, W. L. (1965), *The Age of Equipoise: A Study of the Mid-Victorian Generation* (New York, Norton).

Butler, S. (1879), *Evolution, Old and New; Or, The Theories of Buffon, Dr Erasmus Darwin, and Lamarck, as Compared with that of Mr Charles Darwin* (London, Hardwicke & Bogue).

Cannon, S. F. (1978), *Science in Culture: The Early Victorian Period* (New York, Dawson and Science History Publications).

Cannon, W. F. (1960), "The Problem of Miracles in the 1830s", *Victorian Studies*, 4: 5–32.

—— (1961), "The Bases of Darwin's Achievement: A Revaluation", *Victorian Studies*, 5: 109–134.

—— (1964), "Scientists and Broad Churchmen: An Early Victorian Intellectual Network", *J. British Studies*, 4: 65–88.

—— (1976), "Charles Lyell, Radical Actualism, and Theory", *Brit. J. Hist. Sci.*, 9: 104–120.

Caldwell, W. H. (1887), "The Embryology of Monotremata and Marsupialia – Part I", *Phil. Trans. Roy. Soc.*, 178B: 463–486.

Cardwell, D. S. L. (1972), *The Organisation of Science in England* (London, Heinemann).

Carpenter, W. B. (1851), *Principles of Physiology, General and Comparative* (London, Churchill), 3rd ed.

—— (1854), *Principles of Comparative Physiology* (London, Churchill), 4th ed.

—— (1888), *Nature and Man: Essays Scientific and Philosophical* (London, Kegan Paul).

Chadwick, O. (1975), *The Secularisation of the European Mind in the Nineteenth Century* (Cambridge University Press).

Challinor, J. (1971), *The History of British Geology: A Bibliographic Study* (Newton Abbot, David & Charles).

Chambers, R. (1844), *Vestiges of the Natural History of Creation* (London, Churchill), 2nd ed.

—— (1846), *Explanations: A Sequel to "Vestiges of the Natural History of Creation"* (London, Churchill), 2nd ed.

—— (1884), *Vestiges of the Natural History of Creation* (London, Churchill), 12th ed.

Chant, C. & J. Fauvel (1980), *Darwin to Einstein: Historical Studies on Science and Belief* (Harlow, Longman).

Clark, G. K. (1962), *The Making of Victorian England* (London, Methuen).

Clark, J. W. & T. M. Hughes (1890), *The Life and Letters of the Reverend Adam Sedgwick* (Cambridge University Press), 2 Vols.

Cockshut, A. O. J. (1964), *The Unbelievers* (London, Collins).

Colbert, E. H. (1971), *Men and Dinosaurs: The Search in Field and Laboratory* (Harmondsworth, Penguin).

Cope, E. D. (1867), "An Account of the Extinct Reptiles which approached the Birds", *Proc. Acad, Nat. Sci. Philadelphia*: 234–235.

—— (1870), "On the Hypothesis of Evolution", in Cope (1887), 128–172.

—— (1871), "Synopsis of the Extinct Batrachia, Reptilia and Aves of North America", *Trans. Am. Phil. Soc.*, 14: 1–252.

—— (1878), "Descriptions of Extinct Batrachia and Reptilia from the Permian Formation of Texas", *Proc. Am. Phil. Soc.*, 17: 505–530.

—— (1880), "A Review of the Modern Doctrine of Evolution", in Cope (1887), 215–240.

—— (1882), "Second Contribution to the History of the Vertebrata of the Permian Formation of Texas", *Proc. Am. Phil. Soc.*, 19: 38–58.

—— (1884), "The Relations between the Theromorphous Reptiles and the Monotreme Mammalia", *Proc. Am. Assoc. Adv. Sci.*, 33: 471–482.

—— (1885), "On the Evolution of the Vertebrata, Progressive and Retrogressive", in Cope (1887), 314–349.

—— (1887), *The Origin of the Fittest, Essays on Evolution* (New York, Appleton).

Cornish, C. J. (1904), *Sir William Flower: A Personal Memoir* (London, Macmillan).

Curwen, E. C. (1940), *The Journal of Gideon Mantell, Surgeon and Geologist* (Oxford University Press).

Dana, J. D. (1863), "On Certain Parallel Relations between the Classes of Vertebrates, and on the Bearing of these relations on the Question of the distinctive features of the Reptilian Birds", *Am. J. Sci.*, 315–321.

Darwin, C. (1859), *The Origin of Species by Means of Natural Selection* (London, Murray).

—— (1872), *The Origin of Species by Means of Natural Selection* (London, Murray) 6th ed., With Additions and Corrections.

Darwin, F. (1887), *The Life and Letters of Charles Darwin* (London, Murray), 3 Vols.

—— & A. C. Seward (1903), *More Letters of Charles Darwin* (London, Murray), 2 Vols.

Dawkins, W. B. (1865), "On the Dentition of the Rhinoceros megarhinus", *Nat. Hist. Rev.*, 5: 399–419.

—— (1868), "Darwin on Variation of Animals and Plants", *Edinburgh Review*, 121: 414–450.

—— (1871), "Darwin on the Descent of Man", *Edinburgh Review*, 134: 195–235.

—— (1878), "A Contribution to the History of the Deer of the European Meiocene and Pleiocene Strata", *Quart, J. Geol. Soc.*, 34: 402–419.

Dean, D. R. (1969), "Hitchcock's Dinosaur Tracks", *American Quarterly*, 21: 639–644.

de Beaumont, L. E. (1852), *Notice sur les Systèmes de Montagnes* (Paris, Bertrand), 3 Vols.

de Beer, G. (1964), *Charles Darwin: Evolution by Natural Selection* (New York, Doubleday).

Delair, J. B. & W. A. S. Sarjeant (1975), "The Earliest Discoveries of Dinosaurs", *Isis*, 66: 5–25.

Desmond, A. J. (1979), "Designing the Dinosaur: Richard Owen's Response to Robert Edmond Grant", *Isis*, 70: 224–234.

Duffin, C. (1978), "The Bath Geological Collections: The Importance of

Certain Vertebrate Fossils Collected by Charles Moore", *Geological Curators Newsletter*, 2: 59–67.

Duncan, D. (1908), *The Life and Letters of Herbert Spencer* (London, Methuen).

Duncan, P. M. (1865), "A Description of the Echinodermata from the Strata on the South-Eastern Coast of Arabia, and at Bagh on the Nerbudda, in the Collection of the Geological Society", *Quart. J. Geol. Soc.*, 21: 349–365.

—— (1877), "Anniversary Address", *Quart. J. Geol. Soc.*, 33: 41–88.

Eiseley, L. (1961), *Darwin's Century: Evolution and the Men who Discovered it* (New York, Anchor Books).

—— (1979), *Darwin and the Mysterious Mr X* (New York, Dutton).

Eng, E. (1978), "Thomas Henry Huxley's Understanding of 'Evolution'", *Hist. Sci.*, 16: 291–303.

Etheridge, R. (1867), "On the Stratigraphical Position of *Acanthopholis Horridus* (Huxley)", *Geol. Mag.*, 4: 67–69.

Evans, Joan (1943), *Time and Chance: The Story of Arthur Evans and his Forbears* (London, Longmans).

Evans, John (1865), "On Portions of a Cranium and of a Jaw, in the Slab Containing the Fossil Remains of the Archaeopteryx", *Nat. Hist. Rev.*, 5: 415–421.

—— (1875), "Anniversary Address", *Quart. J. Geol. Soc.*, 31: xxxvii–lxxvi.

—— (1876), "Anniversary Address", *Quart. J. Geol. Soc.*, 32: 47–121.

Eve, A. S. & C. H. Creasey (1945), *Life and Work of John Tyndall* (London, Macmillan).

Falconer, H. (1863), "On the American Fossil Elephant of the Regions bordering the Gulf of Mexico (E. Columbi, Falc.); with General Observations on the Living and Extinct Species", in Falconer (1868), II, 212–291.

—— (1868), *Palaeontological Memoirs* (London, Hardwicke), 2 Vols.

Farber, P. L. (1976), "The Type-Concept in Zoology during the First Half of the Nineteenth Century", *J. Hist. Biol.*, 9: 93–119.

Fichman, M. (1977), "Wallace: Zoogeography and the Problem of Land Bridges", *J. Hist. Biol.*, 10: 45–63.

Fiske, J. (1876), "Darwinism Verified", in Fiske (1902a), 1–30.

—— (1890), "The Doctrine of Evolution: Its Scope and Purport", in Fiske (1902b), 37–60.

—— (1902a), *Darwinism and Other Essays* (Boston, Houghton, Mifflin).

—— (1902b), *A Century of Science and Other Essays* (Boston, Houghton, Mifflin).

—— (1939), *The Personal Letters of John Fiske: A Small Edition Privately Printed for Members of the Bibliophile Society* (Cedar Rapids, Iowa, The Torch Press).

Flett, J. S. (1937), *The First Hundred Years of the Geological Survey of Great Britain* (London, H. M. Stationery Office).

Flower, W. H. (1870), "Introductory Lecture", *Medical Times and Gazette*, 1: 195–200.

—— (1873), "On Palaeontological Evidence of Gradual Modification of Animal Forms", *Proc. Roy. Inst.*, 7: 94–104.

—— (1898), *Essays on Museums and Other Subjects Connected with Natural History* (London, Macmillan).

—— & R. Lydekker (1891), *An Introduction to the Study of Mammals Living and Extinct* (London, A. & C. Black).

Forbes, E. (1844), "Vestiges of the Natural History of Creation", *The Lancet*, II: 265–6.

—— (1853), "Anniversary Address", *Quart. J. Geol. Soc.*, 9: xxviii–xc.

—— (1854), "Anniversary Address", *Quart. J. Geol. Soc.*, 10: xix–lxxxi.

Foster, M. (1869), "Higher and Lower Animals", *Quarterly Review*, 127: 381–400.

—— & E. R. Lankester (1898–1902), *The Scientific Memoirs of Thomas Henry Huxley* (London, Macmillan), 4 Vols.

Gasman, D. (1971), *The Scientific Origins of National Socialism: Social Darwinism in Ernst Haeckel and the German Monist League* (London and New York, Macdonald, American Elsevier).

Gaudry, A. (1862–7), *Animaux Fossiles et Géologie de l'Attique* (Paris, Savy), 2 Vols.

Geikie, A. (1892), "Anniversary Address", *Quart. J. Geol. Soc.*, 48: 38–179.

Geison, G. L. (1969), "The Protoplasmic Theory of Life and the Vitalist-Mechanist Debate", *Isis*, 60: 273–292.

Gerstner, P. A. (1970), "Vertebrate Paleontology, an Early Nineteenth-Century Transatlantic Science", *J. Hist. Biol.*, 3: 137–148.

Ghiselin, M. T. (1969), *The Triumph of the Darwinian Method* (University of California Press).

—— (1980), "The Darwinian Revolution", *Systematic Zoology*, 29: 105–108.

Gibbon, C. (1878), *The Life of George Combe* (London, Macmillan) 2 Vols.

Gillespie, N. C. (1977), "The Duke of Argyll, Evolutionary Anthropology, and the Art of Scientific Controversy", *Isis*, 68: 40–54.

—— (1979), *Charles Darwin and the Problem of Creation* (University of Chicago Press).

Gillispie, C. C. (1959), *Genesis and Geology* (New York, Harper).

Glass, B., O. Temkin, & W. L. Strauss (1968), *Forerunners of Darwin 1745–1859* (Baltimore, The Johns Hopkins University Press).

Glick, T. F. (1974), *The Comparative Reception of Darwinism* (Austin, University of Texas Press).

Godlee, R. J. (1921), "Thomas Wharton Jones", *Brit. J. Ophthalmology*, 93: 145–181.

Goldfuss, G. A. (1847), *Beiträge zur vorweltlichen Fauna des Steinkohlengebirges* (Bonn, Henry & Cohen).

Goodrich, E. S. (1916), "On the Classification of the Reptilia", *Proc. Roy. Soc.* London, 89B: 261–276.

—— (1930), "Edwin Ray Lankester, 1847–1929", *Proc. Roy. Soc.* London, 106B: x–xv.

Gould, S. J. (1971), "Darwin's Retreat", *Science*, 172: 677–678.

—— (1977), *Ontogeny and Phylogeny* (Cambridge, Harvard University Press).

—— (1979), "Agassiz's Later, Private Thoughts on Evolution: His Marginalia in Haeckel's *Natürliche Schöpfungsgeschichte* (1868)", in Schneer (1979), 277–282.

Gray-Smith, R. (1933), *God in the Philosophy of Schelling* (Philadelphia).

Greene, J. C. (1975), "Reflections on the Progress of Darwin Studies", *J. Hist. Biol.*, 8: 243–273.

—— (1981), *Science, Ideology, and World View: Essays in the History of Evolutionary Ideas* (Berkeley, University of California Press).

Gregory, F, (1977), *Scientific Materialism in Nineteenth Century Germany* (Dordrecht, Reidel).

Gregory, J. T. (1979), "North American Vertebrate Paleontology", in Schneer (1979), 305–335.

Gregory, J. W. (1891), "Peter Martin Duncan", *Geol. Mag.*, 8: 332–336.

Gruber, H. E. (1974), *Darwin on Man: A Psychological Study of Scientific Creativity* (New York, Dutton).

Gruber, J. W. (1960), *A Conscience in Conflict: The Life of St George Jackson Mivart* (New York, Columbia University Press).

Gunther, A. E. (1980), *The Founders of Science at the British Museum, 1753–1900* (Suffolk, Halesworth Press).

Guralnick, S. M. (1972), "Geology and Religion Before Darwin: The Case of Edward Hitchcock, Theologian and Geologist (1793–1864)", *Isis*, 63: 529–543.

Haeckel, E. (1866), *Generelle Morphologie der Organismen* (Berlin, Reimer), 2 Vols.

—— (1876), *The History of Creation: Or the Development of the Earth and its Inhabitants by the Action of Natural Causes* (New York, Appleton).

—— (1904), *The Wonders of Life: A Popular Study of Biological Philosophy* (London, Watts).

Haines, G. (1969), *Essays on German Influence upon English Education and Science* (Hamden, Conn., Archon Books).

Hamilton, W. J. (1856), "Anniversary Address", *Quart. J. Geol. Soc.*, 12: xxvi–cxix.

Helfand, M. S. (1977), "T. H. Huxley's 'Evolution and Ethics': The Politics of Evolution and the Evolution of Politics", *Victorian Studies*, 20: 159–177.

Heron-Allen, E. (1929), "Ray Lankester", *J. Roy. Micros. Soc.*, 49: 359–365.

Hitchcock, E. (1848), "An Attempt to Discriminate and Describe the Animals that made the fossil footprints of the United States, and especially in New England", *Mem. Am. Acad. Arts Sci.*, 3: 129–256.

—— (1863), "New Facts and Conclusions respecting the Fossil Footmarks of the Connecticut Valley", *Am. J. Sci.*, 36: 46–57.

Hodge, M. J. S. (1972), "The Universal Gestation of Nature: Chambers' *Vestiges* and *Explanations*", *J. Hist. Biol.*, 5: 127–151.

—— (1974), "England", in Glick (1974), 3–31.

Hofstadter, R. (1955), *Social Darwinism in American Thought* (Boston, Beacon Press).

Howes, G. B. (1905), "St George Mivart", *Proc. Roy. Soc. London*, 75: 95–100.

Huddleston, W. H. (1893), "Anniversary Address", *Quart. J. Geol. Soc.*, 49: 46–142.

Hulke, J. W. (1871), "Note on a Large Reptilian Skull from Brooke, Isle of Wight, probably Dinosaurian, and referable to the Genus Iguanodon", *Quart. J. Geol. Soc.*, 27: 199–206.

—— (1873), "Contribution to the Anatomy of Hypsilophodon Foxii", *Quart. J. Geol. Soc.*, 29: 522–532.

—— (1874a), "Supplemental Note on the Anatomy of Hypsilophodon Foxii", *Quart. J. Geol. Soc.*, 30: 18–23.

—— (1874b), "Note on an Astragalus of Iguanodon Mantelli", *Quart. J. Geol. Soc.*, 30: 24–26.

—— (1876), "Appendix to 'Note on a Modified Form of Dinosaurian Ilium, hitherto reputed Scapula'", *Quart. J. Geol. Soc.*, 32: 364–366.

—— (1878), "Note on Two Skulls from the Wealdon and Purbeck Formations indicating a new Subgroup of Crocodilia", *Quart. J. Geol. Soc.*, 34: 377–382.

—— (1882), "An Attempt at a complete osteology of *Hypsilophodon Foxii*: A British Wealden Dinosaur", *Phil. Trans. Roy. Soc.*, 173: 1035–1062.

Hull, D. L. (1973), *Darwin and His Critics* (Cambridge, Harvard University Press).

Huxley, L. (1900), *Life and Letters of Thomas Henry Huxley* (London, Macmillan), 2 Vols.

—— (1918), *Life and Letters of Sir Joseph Dalton Hooker* (London, Murray), 2 Vols.

Huxley, T. H. (1851a), "Observations upon the Anatomy and Physiology of Salpa and Pyrosoma", in Foster & Lankester (1898–1902), I, 38–68.

—— (1851b), "Report upon the Researches of Prof. Muller into the Anatomy and Development of the Echinoderms", in Foster & Lankester (1898–1902), I, 103–121.

—— (1852), "Upon Animal Individuality", in Foster & Lankester (1898–1902), I, 146–151.

—— (1853), "On the Morphology of the Cephalous Mollusca", in Foster & Lankester (1898–1902), I, 152–193.

—— (1854), "The Vestiges of Creation", *Brit. &. For. Medico-Chirurgical Review*, 26: 425–439.

—— (1854–5), "Contributions to the Anatomy of the Brachiopoda", in Foster & Lankester (1898–1902), I, 325–336.

—— (1855), "Contemporary Literature: Science", *Westminster Review*, 7: 239–253.

—— (1856a), "Owen & Rymer Jones on Comparative Anatomy", *Brit. & For. Medico-Chirurgical Review*, 35: 1–27.

—— (1856b), "On the Method of Palaeontolgy", in Foster & Lankester (1898–1902), I, 432–444.

—— (1858), "On the Theory of the Vertebrate Skull", in Foster & Lankester (1898–1902), I, 538–606.

—— (1859a), "On a Fragment of a Lower Jaw of a Large Labyrinthodont from Cubbingdon", in Foster & Lankester (1898–1902), II, 269–270.

—— (1859b), "The Darwinian Hypothesis", in Huxley (1899), 1–21.

—— (1859c), "On the Persistent Types of Animal Life", in Foster & Lankester (1898–1902), II, 90–93.

—— (1859d), "On the Stagonolepis Robertsoni (Agassiz) of the Elgin Sandstones", in Foster & Lankester (1898–1902), II, 94–119.

—— (1860a), "The Origin of Species", in Huxley (1899), 22–79.

—— (1860b), "On Species and Races, and their Origin", in Foster & Lankester (1898–1902), II, 388–394.

—— (1861), "On the Zoological Relations of Man with the Lower Animals", in Foster & Lankester (1898–1902), II, 471–492.

—— (1862a), "Anniversary Address", in Foster & Lankester (1898–1902), II, 512–529.

—— (1862b), *On Our Knowledge of the Causes of the Phenomena of Organic Nature* (London, Hardwicke).

—— (1862c), "On New Labyrinthodonts from the Edinburgh Coal-Field", in Foster & Lankester (1898–1902), II, 530–535.

—— (1863a), *Evidence as to Man's Place in Nature* (London, Williams & Norgate).

—— (1863b), "Description of Anthracosaurus Russelli, a New Labyrinthodont from the Lanarkshire Coal-Field", in Foster & Lankester (1898–1902), II, 558–572.

—— (1864a), *Lectures on the Elements of Comparative Anatomy* (London, Churchill).

—— (1864b), "Criticisms on 'The Origin of Species'", in Huxley (1899), 80–106.

—— (1865a), "Emancipation – Black and White", in Huxley (1870f), 23–30.

—— (1865b), "Explanatory Preface to the Catalogue of the Palaeontological Collection in the Museum of Practical Geology", in Foster & Lankester (1898–1902), III, 125–175.

—— (1866), "On the Advisableness of Improving Natural Knowledge" in Huxley (1870f), 3–22.

—— (1867a), "On the Classification of Birds", in Foster & Lankester (1898–1902), III, 239–297.

—— (1867b), "On a New Species of Telerpeton Elginense", in Foster & Lankester (1898–1902), III, 205–213.

—— (1868a), "On the Animals which are most nearly intermediate between Birds and Reptiles", in Foster & Lankester (1898–1902), III, 303–313.

—— (1868b), "Remarks upon Archaeopteryx Lithographics", in Foster & Lankester (1898–1902), III, 340–345.

—— (1868c), "On the Classification and Distribution of the Alectomorphae and Heteromorphae", in Foster & Lankester (1898–1902), III, 346–373.

—— (1868d), "On the Physical Basis of Life", in Huxley (1901a), 130–165.

—— (1868e), "A Liberal Education: And Where to Find it", in Huxley (1870f), 31–59.

—— (1868f), "On *Saurosternon Bainii*, and *Pristerodon McKayi*, Two new Fossil Lacertilian Reptiles from South Africa", in Foster & Lankester (1898–1902), III, 298–302.

—— (1869a), "On Hyperodapedon", in Foster & Lankester (1898–1902), III, 374–390.

—— (1869b), "The Natural History of Creation", *The Academy*, 1: 13–14, 40–43.

—— (1869c), "Anniversay Address", in Foster & Lankester (1898–1902), III, 397–426.

—— (1870a), "Anniversary Address", in Foster & Lankester (1898–1902), III, 510–550.

—— (1870b), "On Descartes' 'Discourse Touching the Method of Using One's Reason Rightly and of Seeking Scientific Truth'", in Huxley (1901a), 166–198.

—— (1870c), "On Hypsilophodon Foxii, a new Dinosaurian from the Wealden of the Isle of Wight", in Foster & Lankester (1898–1902), III, 454–464.

—— (1870d), "Further Evidence of the Affinity between the Dinosaurian Reptiles and Birds", in Foster & Lankester (1898–1902), III, 465–486.

—— (1870e), "On the Classification of the Dinosauria with Observations on the Dinosauria of the Trias", in Foster & Lankester (1898–1902), III, 487–509.

—— (1870f), *Lay Sermons, Addresses, and Reviews* (London, Macmillan).

—— (1871a), "Mr Darwin's Critics", in Huxley (1899), 120–186.

—— (1871b), "Administrative Nihilism", in Huxley (1901a), 251–289.

—— (1874a), "On the Hypothesis that Animals are Automata, and its History", in Huxley (1901a), 199–250.

—— (1874b), "Universities: Actual and Ideal", in Huxley (1893b), 189–234.

—— (1875), "On Stagonolepis Robertsoni, and on the Evolution of the Crocodilia", in Foster & Lankester (1898–1902), IV, 66–83.

—— (1876), "On the Evidence as to the Origin of Existing Vertebrate Animals", in Foster & Lankester (1898–1902), IV, 163–187.

—— (1877), *American Addresses* (London, Macmillan).

—— (1879), "On the Characters of the Pelvis in the Mammalia, and the Conclusions respecting the Origin of Mammals which may be based on them", in Foster & Lankester (1898–1902), IV, 345–356.

—— (1880a), "The Coming of Age of the Origin of Species", in Foster & Lankester (1898–1902), IV, 395–403.

—— (1880b), "On the Application of the Laws of Evolution to the Arrangement of the Vertebrata, and more particularly of the Mammalia", in Foster & Lankester (1898–1902), IV, 457–472.

—— (1880c), "On the Method of Zadig", in Huxley (1901b), 1–23.

—— (1887a), "On the Reception of the 'Origin of Species'", in F. Darwin (1887), II, 179–204.

—— (1887b), "Scientific and Pseudo-Scientific Realism", in Huxley (1894c), 59–89.

—— (1887c), "Science and Pseudo Science", in Huxley (1894c), 90–125.

—— (1887d), "The Progress of Science", in Huxley (1901a), 42–129.

—— (1893a), "Autobiography", in Huxley (1901a), 1–17.

—— (1893b), *Science and Education* (London, Macmillan).

—— (1894a), "Professor Tyndall", *Nineteenth Century*, 35: 1–11.

—— (1894b), "Owen's Position in the History of Anatomical Science", in Rev. R. Owen (1894), II, 273–332.

—— (1894c), *Science and Christian Tradition* (London, Macmillan).

—— (1899), *Darwiniana* (London, Macmillan).

—— (1901a), *Method and Results* (London, Macmillan).

—— (1901b), *Science and Hebrew Tradition* (London, Macmillan)

Irvine, W. (1959), *Apes, Angels, and Victorians: Darwin, Huxley, and Evolution* (Cleveland, Meridian).

Jenkin, F. (1867), "The Origin of Species", *North British Review*, 46: 277–318.

Jensen, J. V. (1970), "The X Club: Fraternity of Victorian Scientists", *Brit. J. Hist. Sci.*, 5: 63–72.

Jones, H. F. (1920), *Samuel Butler: A Memoir* (London, Macmillan), 2 Vols.

Jones, G. (1980), *Social Darwinism and English Thought: The Interaction between Biological and Social Theory* (Sussex, Harvester).

Jordanova, L. J. & R. S. Porter (1979), *Images of the Earth: Essays in the Environmental Sciences* (Chalfont St Giles, British Society for the History of Science).

Kelly, A. (1981), *The Descent of Darwin: The Popularization of Darwinism in Germany, 1860–1914* (Chapel Hill, University of North Carolina).

Kennedy, J. G. (1978), *Herbert Spencer* (Boston, Mass., Twayne).

King, C. (1877), "Catastrophism and Evolution", *Am. Nat.*, 11: 449–470.

Knight, D. (1981), *Ordering the World: A History of Classifying Man* (London, Burnett Books).

Kovalevskii, V. (1873), "On the Osteology of the Hyopotamidae", *Phil. Trans. Roy. Soc.*, 163: 19–94.

Lankester, E. R. (1870), "On the Use of the Term Homology in modern Zoology, and the Distinction between Homogenetic and Homoplastic Agreements", *Ann. & Mag. Nat. Hist.*, 6: 34–43.

—— (1877), "Notes on Embryology and Classification of the Animal Kingdom: Comprising a Revision of Speculations Relative to the Origin and Significance of the Germ-Layers", *Quart. J. Micros. Soc.*, 17: 399–454.

—— (1880), *Degeneration: A Chapter in Darwinism* (London, Macmillan).

—— (1883), "Biology and the State", in Lankester (1890), 63–117.

—— (1890), *The Advancement of Science: Occasional Essays and Addresses* (London, Macmillan).

—— (1899), "William Henry Flower", *Nature*, 60: 252–255.

Lartet, E. (1856), "Note sur un grand Singe fossile qui se rattache au groupe des Singes supérieurs", *Comptes Rendus de l'Academie des Sciences*, 43: 219–223.

Lawrence, P. (1978), "Charles Lyell versus the Theory of Central Heat: A Reappraisal of Lyell's Place in the History of Geology", *J. Hist. Biol.*, 11: 101–128.

Levere, T. H. (1981), *Poetry Realized in Nature: Samuel Taylor Coleridge and Early Nineteenth-Century Science* (Cambridge University Press).

Lewes, G. H. (1852), "Goethe as a Man of Science", *Westminster Review*, 58: 479–506.

Lohrli, A. (1973), *Household Words: A Weekly Journal 1850–1859 Conducted by Charles Dickens* (University of Toronto Press).

Lovejoy, A. O. (1968), "The Argument for Organic Evolution before the *Origin of Species*, 1830–1858", in Glass *et al* (1968), 356–414.

Lucas, J. R. (1979), "Wilberforce and Huxley: A Legendary Encounter", *Historical Journal*, 22: 313–330.

Lydekker, R. (1906), *Sir William Flower* (London, Dent).

Lyell, C. (1851), "Anniversary Address", *Quart. J. Geol. Soc.*, 7: xxv–lxxvi.

—— (1852), *A Manual of Elementary Geology* (London, Murray), 4th ed.

—— (1859), *Supplement to the Fifth Edition of a Manual of Elementary Geology* (London, Murray).

—— (1863), *The Geological Evidences of the Antiquity of Man* (London, Murray).

—— (1868), *Principles of Geology* (London, Murray), 10th ed., 2 Vols.

Lyell, K. (1881), *Life, Letters and Journals of Sir Charles Lyell* (London, Murray), 2 Vols.

Mackie, S. J. (1863), "The Aeronauts of the Solenhofen Age", *The Geologist*, 6: 1–8.

MacLeod, R. M. (1965), "Evolutionism and Richard Owen, 1830–1868: An Episode in Darwin's Century", *Isis*, 56: 259–280.

—— (1970), "The X-Club: A Social Network of Science in Late-Victorian England", *Notes and Records of the Royal Society*, 24: 305–322.

—— (1971), "The Support of Victorian Science: The Endowment of Research Movement in Great Britain, 1868–1900", *Minerva*, 9: 197–230.

—— (1974), "The Ayrton Incident: A Commentary on the Relations of Science and Government in England, 1870–1873", in Thackray & Mendelsohn (1974), 45–78.

Mandelbaum, M. (1971), *History, Man, and Reason: A Study in Nineteenth-Century Thought* (Baltimore, The Johns Hopkins University Press).

Manier, E. (1978), *The Young Darwin and His Cultural Circle* (Dordrecht, Reidel).

Marchant, J. (1916), *Alfred Russel Wallace: Letters and Reminiscences* (London, Cassell), 2 Vols.

Marsh, O. C. (1874), "Notice of New Equine Mammals from the Tertiary Formation", *Am. J. Sci.*, 7: 247–258.

—— (1877), *Introduction and Succession of Vertebrate Life in America: An Address Delivered before the American Association for the Advancement of Science, at Nashville, Tenn., Aug. 30, 1877*, offprint pp. 1–50.

—— (1878), "Fossil Mammal from the Jurassic of the Rocky Mountains", *Am. J. Sci.*, 15: 459.

—— (1879), *History and Methods of Paleontological Discovery: An Address*

Delivered before the AAAS at Saratoga, N.Y., Aug. 28, 1879, offprint pp. 1–44.

—— (1880), Notice of Jurassic Mammals representing Two New Orders", *Am. J. Sci.*, 20: 235–239.

—— (1882), *Evolution, Once an Hypothesis, now an Established Doctrine of the Scientific World: Address at the Farewell Dinner to Herbert Spencer, in New York, November 9, 1882*, offprint pp. 1–4.

—— (1887), "American Jurassic Mammals", *Am. J. Sci.*, 33: 327–348.

Marx, K. & F. Engels (1943), *Selected Correspondence, 1846–1895* (London, Lawrence & Wishart).

McCoy, F. (1862), "Note on the Ancient and Recent Natural History of Victoria", *Ann. & Mag. Nat. Hist.*, 9: 137–150.

Mendelsohn, E., P. Weingart, & R. Whitley (1977), *The Social Production of Scientific Knowledge* (Dordrecht, Reidel).

Meyer, H. von (1848), "The Reptiles of the Coal Formation", *Quart. J. Geol. Soc.*, 4: Pt. 2, 51–56.

—— (1856–8), "Reptilien aus der Steinkohlen-Formation in Deutschland", *Palaeontographica*, 6: 59–220.

—— (1862), "On the *Archaeopteryx lithographica*, from the Lithographic Slate of Solenhofen", *Ann. & Mag. Nat. Hist.*, 9: 366–370.

Miller, H. (1857), *The Testimony of the Rocks* (Edinburgh, Constable).

Millhauser, M. (1959), *Just Before Darwin: Robert Chambers and Vestiges* (Middletown, Conn., Wesleyan University Press).

Mitchell, P. C. (1929), "Sir E. Ray Lankester", *The Times*, 16 Aug., 15.

Mivart, St G. J. (1869), "Difficulties of the Theory of Natural Selection", *The Month*, 11: 35–53, 134–153, 274–289.

—— (1870), "On the Use of the Term 'Homology'", *Ann. & Mag. Nat. Hist.*, 6: 113–121.

—— (1871), *On the Genesis of Species* (London, Macmillan).

—— (1872), "Specific Genesis", in Mivart (1892), II, 103–126.

—— (1873), "On *Lepilemur* and *Cheirogaleus*, and the Zoological Rank of the Lemuroidea", *Proc. Zool. Soc.*, 41: 485–510.

—— (1875), "Likenesses; or Philosophical Anatomy", in Mivart (1892), II, 250–278.

—— (1878), "Force, Energy, and Will", in Mivart (1892), II, 226–249.

—— (1888), "On the Possibly Dual Origin of the Mammalia", *Proc. Roy. Soc.* London, 43: 372–9.

—— (1892), *Essays and Criticisms* (London, Osgood), 2 Vols.

—— (1893), "Sir Richard Owen's Hypotheses", *Natural Science*, 2: 18–23.

—— (1897), "Some Reminiscences of Thomas Henry Huxley", *Nineteenth Century*, 42: 985–998.

Montgomery, W. M. (1974), "Germany", in Glick (1974), 81–116.

Moore, J. R. (1979), *The Post-Darwinian Controversies: A Study of the Protestant Struggle to Come to Terms with Darwin in Great Britain and America* (Cambridge University Press).

Moyal, A. M. (1975), "Sir Richard Owen and his Influence on Australian Zoological and Palaeontological Science", *Records of the Australian Academy of Sciences*, 3: 41–56.

Mozley, A. (1967), "Evolution and the Climate of Opinion in Australia, 1840–1876", *Victorian Studies*, 10: 411–430.

Mozley, J. B. (1869), "The Argument of Design", *Quarterly Review*, 127: 134–176.

Mulkay, M. (1979), *Science and the Sociology of Knowledge* (London, George Allen & Unwin).

Murchison, R. I. (1864), "Address", *J. Roy. Geog. Soc.*, 34: cix–cxciii.

Murphy, H. R. (1955), "The Ethical Revolt against Christian Orthodoxy in Early Victorian England", *American Historical Review*, 60: 800–817.

Murray, A. (1866), *The Geographical Distribution of Mammals* (London, Day).

Napier, M. (1879), *Selection from the Correspondence of the Late Macvey Napier, Esq.* (London, Macmillan).

Nelson, G. (1978), "From Candolle to Croizat: Comments on the History of Biogeography", *J. Hist. Biol.*, 11: 269–305.

Neve, M. (1980), "The Naturalization of Science", *Social Studies of Science*, 10: 375–391.

Nordenskiöld, E. (1942), *The History of Biology: A Survey* (New York, Tudor).

Oldroyd, D. R. (1980a), *Darwinian Impacts* (Milton Keynes, Open University Press).

—— (1980b), "Sir Archibald Geikie (1835–1924), Geologist, Romantic Aesthete, and Historian of Geology: The Problem of Whig Historiography of Science", *Annals of Science*, 37; 441–462.

Oliver, D. (1862), "The Atlantis Hypothesis in its Botanical Aspect", *Natural History Review*, 2: 149–170.

Oppenheimer, J. M. (1967), *Essays in the History of Embryology and Biology* (Cambridge, MIT Press).

Osborn, H. F. (1910), *The Age of Mammals* (New York, Macmillan).

—— (1931), *Cope: Master Naturalist. The Life and Letters of Edward Drinker Cope* (Princeton University Press).

Ospovat, D. (1976), "The Influence of Karl Ernst von Baer's Embryology, 1828–1859: A Reappraisal in light of Richard Owen's and William B. Carpenter's 'Palaeontological Application of von Baer's Law'", *J. Hist. Biol.*, 9: 1–28.

—— (1977), "Lyell's Theory of Climate", *J. Hist. Biol.*, 10: 317–339.

—— (1978), "Perfect Adaptation and Teleological Explanation: Approaches to the Problem of the History of Life in the Mid-Nineteenth Century", *Studies in the History of Biology*, 2: 33–56.

—— (1980), "God and Natural Selection: The Darwinian Idea of Design", *J. Hist. Biol.*, 13: 169–194.

—— (1981), *The Development of Darwin's Theory: Natural History, Natural Theology, and Natural Selection, 1838–1859* (Cambridge University Press).

Owen, Rev. R. (1894), *The Life of Richard Owen* (London, Murray), 2 Vols.

Owen, R. (1835), "On the Osteology of the Chimpanzee and Orang Utan", *Trans. Zool. Soc.* London, 1: 343–379.

—— (1841), "Report on British Fossil Reptiles: Part II", *Report of the British Association for the Advancement of Science* (Plymouth Meeting), 60–204.

—— (1846a), *A History of British Fossil Mammals and Birds* (London, van Voorst).

—— (1846b), "Report on the Archetype and Homologies of the Vertebrate Skeleton", *Report of the British Association for the Advancement of Science*, (Southampton Meeting), 169–340.

—— (1849a), *On the Nature of Limbs* (London, van Voorst).

—— (1849b), "Osteological Contributions to the Natural History of the Chimpanzees (*Troglodytes*, Geoffroy), including the Descriptions of the skull of a large species (*Troglodytes* Gorilla, Savage) discovered by Thomas S. Savage, MD in the Gaboon Country, West Africa", *Trans. Zool. Soc.* London, 3: 381–422.

—— (1849c), *On Parthenogenesis, or the Successive Production of Procreating Individuals from a Single Ovum* (London, van Voorst).

—— (1851), "Lyell – on Life and Successive Development", *Quarterly Review*, 89; 412–451.

?—— (1852), "Dr Mantell", *Literary Gazette*, No. 1869: 842.

—— (1854a), *Geology and Inhabitants of the Ancient World* (London, Bradbury & Evans).

—— (1854b), *Descriptive Catalogue of the Fossil Organic Remains of Reptilia and Pisces contained in the Museum of the Royal College of Surgeons of England* (London, Taylor & Francis).

—— (1854c), "On Metamorphosis and Metagenesis", *Proc. Roy. Inst.*, 1: 9–16.

—— (1855), *Lectures on the Comparative Anatomy and Physiology of the Invertebrate Animals* (London, Longman).

—— (1857), "On the Affinities of the *Stereognathus Ooliticus* (Charlesworth), a Mammal from the Oolitic Slate of Stonesfield", *Quart. J. Geol. Soc.*, 13: 1–11.

—— (1858a), "Presidential Address", *Report of the British Association for the Advancement of Science* (Leeds Meeting), xlix–cx.

—— (1858b), "On the Characters, Principles of Division, and Primary Groups of the Class Mammalia", *J. Proc. Linn. Soc.* (Zool.), 2: 1–37.

—— (1859a), *On the Classification and Distribution of the Mammalia* (London, Parker).

—— (1859b), "On the Orders of Fossil and Recent Reptilia, and their Distribution in Time", *Report of the British Association for the Advancement of Science* (Aberdeen Meeting), 153–166.

—— (1859c), "Palaeontology", *Encyclopaedia Britannica*, 8th ed., 17: 91–176.

—— (1860a), *Palaeontology, or A Systematic Summary of Extinct Animals and their Geological Relations* (Edinburgh, Black).

—— (1860b), "Darwin on the Origin of Species", *Edinburgh Review*, 111: 487–532.

—— (1860c), "On Some Reptilian Remains from South Africa", *Quart. J. Geol. Soc.*, 16: 49–63.

—— (1861a), *Palaeontology* (Edinburgh, Black), 2nd ed.

—— (1861b), "The Gorilla and the Negro", *Athenaeum*, 23 March: 395–6.

—— (1862), "On the Dicynodont Reptilia, with a Description of some Fossil Remains brought by HRH Prince Alfred from South Africa, November 1860", *Phil. Trans. Roy. Soc.*, 152: 455–467.

—— (1863a), "On the Archaeopteryx of von Meyer, with a Description of the Fossil Remains of a Long-Tailed Species, from the Lithographic Slate of Solenhofen", *Phil. Trans. Roy. Soc.*, 153: 33–47.

—— (1863b), "Ape-Origin of Man as Tested by the Brain", *Athenaeum*, 21 February: 262–263.

—— (1863c), *Monograph on the Aye-Aye* (London, Taylor & Francis).

—— (1864), "Instances of the Power of God as Manifested in His Animal Creation", in *Lectures delivered before the Young Men's Christian Association* (London, Simpkin & Marshall), 1–35.

—— (1866–8), *On the Anatomy of Vertebrates* (London, Longman), 3 Vols.

—— (1869), *Monograph of the Fossil Reptilia of the Liassic Formations* (London, Palaeontographical Society).

—— (1871), *Monograph of the Fossil Mammalia of the Mesozoic Formation* (London, Palaeontographical Society).

—— (1872), "The Fate of the 'Jardin d'Acclimatation' during the late siege of Paris", *Fraser's Magazine*, January: 17–22.

—— (1872–1889), *A Monograph on the Fossil Reptilia of the Wealden and Purbeck Formations* (London, Palaeontographical Society).

—— (1874–1889), *Monograph on the Fossil Reptilia of the Mesozoic Formations* (London, Palaeontographical Society).

—— (1876a), *Descriptive and Illustrated Catalogue of the Fossil Reptilia of South Africa* (London, Taylor & Francis).

—— (1876b), "Evidence of a Carnivorous Reptile (Cynodraco Major, Ow.) about the Size of a Lion, with remarks thereon", *Quart. J. Geol. Soc.*, 32: 95–101.

—— (1876c), "Evidence of Theriodonts in Permian Deposits elsewhere than in South Africa", *Quart. J. Geol. Soc.*, 32: 352–363.

—— (1876d), "On a new Modification of Dinosaurian Vertebrae", *Quart. J. Geol. Soc.*, 32: 43–46.

—— (1878), "On the Influence of the Advent of a Higher Form of Life in Modifying the Structure of an Older and Lower Form", *Quart. J. Geol. Soc.*, 34: 421–430.

—— (1879), "On the Association of Dwarf Crocodiles (*Nannosuchus* and *Theriosuchus pusillus*, e.g.) with the Diminutive Mammals of the Purbeck Shales", *Quart. J. Geol. Soc.*, 35: 148–155.

—— (1880), "Description of Parts of the Skeleton of an Anomodont Reptile (Platypodosaurus Robustus, Ow.) from the Trias of Graaf Reinet, S. Africa", *Quart. J. Geol. Soc.*, 36: 414–425.

—— (1881), "On the Order Theriodontia, with a Description of a new Genus and Species (Aelurosaurus Felinus, Ow.)", *Quart. J. Geol. Soc.*, 37: 261–265.

—— & J. Broderip (1853), "Generalizations of Comparative Anatomy", *Quarterly Review*, 93: 46–83.

Paradis, J. G. (1978), *T. H. Huxley: Man's Place in Nature* (Lincoln, University of Nebraska Press).

Parker, W. K. (1864a), "Remarks on the Skeleton of the Archaeopteryx; and on the Relations of the Bird to the Reptile", *Geol. Mag.*, 1:55–57.

—— (1864b), "On the Sternal Apparatus of Birds and Other Vertebrata", *Proc. Zool. Soc.*, 339–341.

Parker, T. J. (1893), *William Kitchen Parker* (London, Macmillan).

Peckham, M. (1959), "Darwinism and Darwinisticism", in Peckham (1970), 176–201.

—— (1970, *The Triumph of Romanticism* (Columbia, University of South Carolina Press).

Peel, J. D. Y. (1971), *Herbert Spencer: The Evolution of a Sociologist* (London, Heinemann).

Pfeifer, E. J. (1974), "United States", in Glick (1974), 168–206.

Phillips, J. (1860a), *Life on the Earth: Its Origin and Succession* (Cambridge, Macmillan).

—— (1860b), "Anniversary Address", *Quart. J. Geol. Soc.*, 16: xxvii–lv.

—— (1871), *Geology of Oxford and the Valley of the Thames* (Oxford, Clarendon).

Poore, G. V. (1901), "Robert Edmond Grant", *University College Gazette*, 2 (34): 190–191.

Porter, R. (1976), "Charles Lyell and the Principles of the History of Geology", *Brit. J. Hist. Sic.*, 9: 91–103.

—— (1977), *The Making of Geology: Earth Science in Britain, 1660–1815* (Cambridge University Press).

—— (1978), "Gentlemen and Geology: The Emergence of a Scientific Career, 1660–1920", *The Historical Journal*, 21: 809–836.

Portlock, J. E. (1858), "Anniversary Address", *Quart. J. Geol. Soc.*, 14: xxi–clxiii.

Poulton, E. B. (1888), "True Teeth in the Young *Ornithorhynchus paradoxus*", *Proc. Roy. Soc.*, 43: 353–356.

—— (1888–9), "The True Teeth and the Horny Plates of *Ornithorhynchus*", *Quart. J. Micros. Soc.*, 29: 9–48.

—— (1908a), "Thomas Henry Huxley and the Theory of Natural Selection", in Poulton (1908b), 193–219.

—— (1908b), *Essays on Evolution 1889–1907* (Oxford, Clarendon Press).

Powell, B. (1838), *The Connexion of Natural and Divine Truth* (London, Parker).

—— (1856), *The Unity of Worlds* (London, Longman), 2nd ed. (First edition title: *Essays on the Spirit of the Inductive Philosophy, the Unity of Worlds and the Philosophy of Creation.*)

—— (1859), *The Order of Nature, Considered in Reference to the Claims of Revelation* (London, Longman).

Rainger, R. (1981), "The Continuation of the Morphological Tradition: American Paleontology, 1880–1910", *J. Hist. Biol.*, 14: 129–158.

Rehbock, P. F. (1978), "*Fossils and Progress*", *Archives Internat. Hist. Sciences*, 28: 340–342.

Richardson, R. A. (1981), "Biogeography and the Genesis of Darwin's Ideas on Transmutation", *J. Hist. Biol.*, 14: 1–41.

Ridley, M. (1982), "Coadaptation and the Inadequacy of Natural Selection", *Brit. J. Hist. Sci.*, 15: 45–68.

Rinard, R. G. (1981), "The Problem of the Organic Individual: Ernst Haeckel and the Development of the Biogenetic Law", *J. Hist. Biol.*, 14: 249–275.

Roderick, G. W. & M. D. Stephens (1972), *Scientific and Technical Education in Nineteenth-Century England* (New York, Barnes & Noble).

Rogers, J. A. (1973), "The Reception of Darwin's *Origin of Species* by Russian Scientists", *Isis*, 64: 484–503.

Rolleston, G. (1861), "On Correlations of Growth, with a Special Example from The Anatomy of a Porpoise", in Rolleston (1884), I, 56–61.

—— (1884), *Scientific Papers and Addresses* (Oxford, Clarendon), 2 Vols.

Rorison, G. (1862), "The Creative Week", in *Replies to "Essays and*

Reviews" (Oxford and London, Henry & Parker), 2nd ed., 277–345.

Rothblatt, S. (1968), *The Revolution of the Dons: Cambridge and Society in Victorian England* (London, Faber).

Rudwick, M. J. S. (1970), "The Strategy of Lyell's *Principles of Geology*", *Isis*, 61: 5–33.

—— (1972), *The Meaning of Fossils: Episodes in the History of Palaeontology* (London, Macdonald).

—— (1978a), "Review – *Fossils and Progress*", *Am. J. Sci.*, 278: 95–96.

—— (1978b), "Transposed Concepts from the Human Sciences in the Early Work of Charles Lyell", in Jordanova & Porter (1978), 67–83.

—— (1980), "Social Order and the Natural World", *Hist. Sci.*, 18: 269–285.

Ruse, M. (1979), *The Darwinian Revolution: Science Red in Tooth and Claw* (Chicago University Press).

Russell, E. S. (1916), *Form and Function: A Contribution to the History of Animal Morphology* (London, Murray).

Russett, C. E. (1976), *Darwin in America: The Intellectual Response, 1865–1912* (San Francisco, W. H. Freeman).

Rütimeyer, L. (1867), *Ueber die Herkunft Unsurer Thierwelt. Eine Zoogeographische Skizze* (Basel & Geneva, H. Georg).

—— (1867–8), "Versuch einer natürlichen Geschichte des Rindes, in seinen Beziehungen zu den Wiederkauern im Allgemeinen", *Neue Denkschriften*, Tom. 22 (1867), offprint pp. 1–102; Tom. 23 (1868), offprint pp. 1–175.

—— (1868), *Die Grenzen der Thierwelt. Eine Betrachtung zu Darwin's Lehre* (Basel, H. Richter).

Sarjeant, W. A. S. (1980), *Geologists and the History of Geology* (London, Macmillan), 5 Vols.

Schäfer, E. A. (1901), "William Sharpey", *University College Gazette*, 2(36): 214–215.

Schelling, F. W. J. (1966), *On University Studies* (Athens, Ohio University Press).

Schmidt, O. (1874), *The Doctrine of Descent and Darwinism* (New York, Appleton).

Schneer, C. J. (1969), *Toward a History of Geology* (Cambridge, MIT Press).

—— (1979), *Two Hundred Years of Geology in America* (Hanover, University Press of New England).

Schuchert, C. & C. M. LeVene (1940), *O. C. Marsh: Pioneer in Paleontology* (New Haven, Yale University Press).

Sedgwick, A. (1845), "Vestiges of the Natural History of Creation", *Edinburgh Review*, 82: 1–85.

—— (1860), "Objections to Mr Darwin's Theory of the Origin of

Species", *The Spectator*, 24 March, 7 April: 285–286, 334–335.

Seeley, H. G. (1861), "On Learning Geology", *The Working Men's College Magazine*, offprint pp. 1–4.

—— (1864), "On the Pterodactyle as Evidence of a new Subclass of Vertebrata (Saurornia)", *Report of the British Association for the Advancement of Science* (Bath Meeting), 69.

—— (1866a), "An Epitome of the Evidence that Pterodactyles are not Reptiles, but a New Subclass of Vertebrated Animals Allied to Birds", *Ann. & Mag. Nat. Hist.*, 17: 321–331.

—— (1866b), "Outline of a Theory of the Skull and the Skeleton", *Ann. & Mag. Nat. Hist.*, 18: 345–362.

—— (1867), "Theory of the Skull and the Skeleton", *Ann. & Mag. Nat. Hist.*, 19: 371–372.

—— (1870a), "Remarks on Prof. Owen's Monograph on Dimorphodon", *Ann. & Mag. Nat. Hist.*, 6: 129–152.

—— (1870b), *The Ornithosauria* (Cambridge, Deighton, Bell).

—— (1871), "On Acantholis platypus (Seeley), a Pachypod from the Cambridge Upper Greensand", *Ann. & Mag. Nat. Hist.*, 8: 305–318.

—— (1872), "The Origin of the Vertebrate Skeleton", *Ann. & Mag. Nat. Hist.*, 10: 21–45.

—— (1875a), "On the Maxillary Bone of a new Dinosaur (Priodontognathus Phillipsii), contained in the Woodwardian Museum of the University of Cambridge", *Quart. J. Geol. Soc.*, 31: 439–443.

—— (1875b), "On the Femur of Cryptosaurus Eumerus, Seeley, a Dinosaur from the Oxford Clay of Great Gransden", *Quart. J. Geol. Soc.*, 31: 149–151.

—— (1876), "On the Organization of the Ornithosauria", *Zool. J. Linn. Soc.*, 13: 84–107.

—— (1879), "Note on a Femur and a Humerus of a Small Mammal from the Stonesfield Slate", *Quart. J. Geol. Soc.*, 35: 456–463.

—— (1882), *The History of the Skull* (Wallington, W. Pile).

—— (1887), "On the Classification of the Fossil Animals Commonly Named Dinosauria", *Proc. Roy. Soc.*, 43: 165–171.

—— (1889), "Some Scientific Results of a Mission to South Africa", *Trans. S. African Phil. Soc.*, 1–16.

—— (1891), "The Ornithosaurian Pelvis", *Ann. & Mag. Nat. Hist.*, 7: 237–255.

—— (1901), *Dragons of the Air* (London, Methuen).

Shairp, J. C., P. G. Tait, & A. Adams-Reilly (1873), *Life and Letters of James David Forbes* (London, Macmillan).

Shapin, S. (1982), "History of Science and its Sociological Reconstructions", *History of Science*, 20: in press.

Sharlin, H. I. (1979), *Lord Kelvin: The Dynamic Victorian* (Pennsylvania State University Press).

Sharpey, W. (1862), "The Address in Physiology", *Brit. Med. J.*, ii: 162–171.

Simpson, G. G. (1951), *Horses* (New York, Oxford University Press).

Smith, R. (1972), "Alfred Russel Wallace: Philosophy of Nature and Man", *Brit. J. Hist. Sci.*, 6: 177–199.

—— (1973), "The Background of Physiological Psychology in Natural Philosophy", *Hist. Sci.*, 11: 75–123.

—— (1974), "Essay Review: Darwin and His Critics", *Brit. J. Hist. Sci.*, 7: 278–285.

Smith, C. U. M. (1982), "Evolution and the Problem of Mind: Part I. Herbert Spencer", *J. Hist. Biol.*, 15: 55–88.

Sollas, W. J. (1909), "Anniversary Address", *Quart. J. Geol. Soc.*, 65: l–cxxii.

Spencer, H. (1857), "Progress: Its Law and Cause", *Westminster Review*, 67: 445–485.

—— (1858), "Owen on the Homologies of the Vertebrate Skeleton", *Brit. & For. Medico-Chirurgical Review*, 44: 400–416.

—— (1859), "Illogical Geology", in Spencer (1868), I: 325–376.

—— (1862), *First Principles* (London, Williams & Norgate).

—— (1864), *The Principles of Biology* (London, Williams & Norgate), 2 Vols.

—— (1868), *Essays: Scientific, Political, and Speculative* (London, Williams & Norgate).

—— (1904), *An Autobiography* (London, Williams & Norgate), 2 Vols.

Stauffer, R. C. (1957), "Haeckel, Darwin, and Ethology", *Quart. Rev. Biol.*, 32: 138–144.

—— (1975), *Charles Darwin's Natural Selection, Being the Second Part of his Big Species Book Written from 1856 to 1858* (Cambridge University Press).

Stearn, W. T. (1981), *The Natural History Museum at South Kensington: A History of the British Museum (Natural History) 1753–1980* (London, Heinemann).

Stebbins, R. E. (1974), "France", in Glick (1974), 117–163.

Swinton, W. E. (1962), "Henry Govier Seeley and the Karoo Reptiles", *Bull. Brit. Mus. (Nat. Hist.) Historical Series*, 3: 1–39.

Tait, P. G. (1869), "Geological Time", *North British Review*, 11:406–439.

Taylor, D. W. (1971), "The Life and Teaching of William Sharpey (1802–1880) 'Father of Modern Physiology' in Britain", *Medical History*, 15: 126–153, 241–259.

Teich, M. & R. M. Young (1973), *Changing Perspectives in the History of Science* (London, Heinemann).

Temkin, C. (1968), "The Idea of Descent in Post-Romantic German Biology, 1848–1858", in Glass *et al* (1968), 323–355.

Thackray, A., and E. Mendelsohn (1974), *Science and Values: Patterns of Tradition and Change* (New York, Humanities Press).

Thomas, O. (1888), *Catalogue of the Marsupialia and Monotremata in the Collections of the British Museum (Natural History)* (London, British Museum [Natural History]).

Thomson, W. (1868), "On Geological Time", in Thomson (1889–1894), 10–64.

—— (1889–1894), *Popular Lectures and Addresses* (London, Macmillan), 3 Vols.

Todes, D. P. (1978), "V. O. Kovalevskii: The Genesis, Content, and Reception of his Paleontological Work", *Studies in the History of Biology*, 2: 99–165.

Turner, F. M. (1974), *Between Science and Religion: The Reaction to Scientific Naturalism in Late Victorian England* (New Haven, Yale University Press).

—— (1978), "The Victorian Conflict between Science and Religion: A Professional Dimension", *Isis*, 69: 356–376.

—— (1980), "Public Science in Britain, 1880–1919", *Isis*, 71: 589–608.

Tyndall, J. (1863), "Vitality", in Tyndall (1871), 436–444.

—— (1867), "Matter and Force", in Tyndall (1871), 71–94.

—— (1870), "On the Scientific Use of the Imagination", in Tyndall (1871), 127–167.

—— (1871), *Fragments of Science* (London, Longman), 2nd ed.

—— (1874), "Belfast Address", in Tyndall (1879), 137–203.

—— (1877), "Science and Man", in Tyndall (1879), 337–374.

—— (1879), *Fragments of Science II* (London, Longman) 6th ed.

Unger, F. (1865a), "The Sunken Island of Atlantis", *J. Botany*, 3: 12–26.

—— (1865b), "New Holland in Europe", *J. Botany*, 3: 39–70.

Vogt, C. (1864), *Lectures on Man: His Place in Creation, and in the History of the Earth* (London, Longman).

Vorzimmer, P. J. (1972), *Charles Darwin: The Years of Controversy* (University of London Press).

Vucinich, A. (1974), "Russia: Biological Sciences", in Glick (1974), 227–255.

Wagner, A. (1862), "On a new Fossil Reptile supposed to be furnished with Feathers", *Ann. & Mag. Nat. Hist.*, 9: 261–267.

Wagner, R. (1861), "On the Structure of the Brain in Man and the Apes, and its Relation to the Zoological System", *Ann. & Mag. Nat. Hist.*, 8: 429–431.

Wallace, A. R. (1876), *The Geographical Distribution of Animals* (London, Macmillan), 2 Vols.

Whewell, W. (1840), *The Philosophy of the Inductive Sciences* (London, Parker), 2 Vols.

—— (1854), *On the Plurality of Worlds* (London, Parker), 2nd ed.

White, A. D. (1965), *A History of the Warfare of Science with Theology in Christendom* (New York, Free Press).

Wilson, L. G. (1970), *Sir Charles Lyell's Scientific Journals on the Species Question* (New Haven, Yale University Press).

Wiltshire, D. (1978), *The Social and Political Thought of Herbert Spencer* (Oxford University Press).

Winsor, M. P. (1976), *Starfish, Jellyfish, and the Order of Life: Issues in Nineteenth-Century Science* (New Haven, Yale University Press).

Wood, P. (1980), "The Dehumanisation of History", *Brit. Soc. Hist. Sci. Newsletter*, 3: 19–20.

Woodward, A. S. (1910), "Harry Govier Seeley", *Proc. Roy. Soc.*, 83B: xv–xvii.

—— (1930–1), "Sir William Boyd Dawkins", *Proc. Roy. Soc.*, 107B: xxiii–xxvi.

Woodward, H. (1862), "On a Feathered Fossil from the Lithographic Limestone of Solenhofen", *Intellectual Observer*, 2: 313–319.

—— (1874), "New Facts bearing on the inquiry concerning forms intermediate between Birds and Reptiles", *Quart. J. Geol. Soc.*, 30: 8–15.

—— (1893), "Sir Richard Owen", *Geol. Mag.*, 10: 49–54.

—— (1896), "Anniversary Address", *Quart. J. Geol. Soc.*, 52: lii–cxviii.

—— (1905), "Sir Frederick McCoy, 1823–1899", *Proc. Roy. Soc.*, 75: 43–45.

Wynne, B. (1979), "Physics and Psychics: Science, Symbolic Action and Social Control in Late Victorian England", in Barnes & Shapin (1979), 167–184.

Young, G. M. (1977), *Portrait of an Age: Victorian England* (London, Oxford University Press), annotated edition.

Young, R. M. (1965), "The Development of Herbert Spencer's Concept of Evolution", *Congrès International d'Histoire des Sciences*, 11 (1), 273–278.

—— (1973), "The Historiographic and Ideological Contexts of the Nineteenth Century Debate on Man's Place in Nature", in Teich & Young (1973), 344–438.

—— (1979), "Interpreting the Production of Science", *New Scientist*, 81: 1026–1028.

—— (1980), "Natural Theology, Victorian Periodicals and the Fragmentation of a Common Context", in Chant & Fauvel (1980), 69–107.

Index